T0406191

Climate Politics and the Impact of Think Tanks

Alexander Ruser

Climate Politics and the Impact of Think Tanks

Scientific Expertise in Germany and the US

palgrave
macmillan

Alexander Ruser
Zeppelin University
Friedrichshafen, Germany

ISBN 978-3-319-75749-0 ISBN 978-3-319-75750-6 (eBook)
https://doi.org/10.1007/978-3-319-75750-6

Library of Congress Control Number: 2018935411

Cover illustration: Gale S. Hanratty / Alamy Stock Photo

Printed on acid-free paper

This Palgrave Macmillan imprint is published by the registered company Springer International Publishing AG part of Springer Nature.
The registered company address is: Gewerbestrasse 11, 6330 Cham, Switzerland

Contents

List of Figures

List of Tables

CHAPTER 1

Introduction

In April 2016 Bill Nye, American TV personality, trusted expert, and self-declared "Science Guy", triggered a small scandal. According to *The Washington Times*, Nye had proposed that climate change dissent was made a criminal, even jailable, offence (Richardson 2016). "Was it appropriate to jail the guys from Enron?" Nye was quoted as saying, continuing "Was it appropriate to jail people from the cigarette industry who insisted that this addictive product was not addictive, and so on?" (Richardson 2016). His words raise some provocative questions. How should societies respond to individuals, groups, and industries that query widely held scientific opinions? Do those who do so deliberately mislead the public? Is it appropriate to compare climate change scepticism and denial with the deliberate and scandalous deception of stakeholder and regulatory authorities by Enron executives, a deception that ultimately led to the downfall of that once powerful corporation and the loss of jobs and pensions of thousands of its employees?

Nye's comparison of climate denial and cigarette industry campaigns is perhaps less controversial; both involve the blatant refutation of scientific evidence. Should, then, the protracted and sophisticated attempts of the tobacco industry and its imaginative, scrupulous, lobbyists be held *legally* accountable for denying the dangers of smoking? When, for instance, the Tobacco Institute in the early 1970s distributed *Smoking and Health: The Need to Know*, a "documentary" which successfully dispelled fears of contracting lung cancer from cigarettes (Proctor 2012: 89), should they have been prosecuted?

A. Ruser, *Climate Politics and the Impact of Think Tanks*,
https://doi.org/10.1007/978-3-319-75750-6_1

Likewise, when the Competitive Enterprise Institute (CEI) launched a costly TV ad campaign entitled "We call it life", asserting the importance of carbon dioxide for plant photosynthesis ("We breathe it out, plants breathe it in"), should it have been considered as engaging in *criminal* activity for distracting from the globally held and scientifically robust claim that carbon dioxide is one of the greenhouse gases driving human-made global climate change? The timing of the campaign reveals its strategic nature. The CEI launched "We call it life" at precisely the time the former US Vice President Al Gore's film *An Inconvenient Truth* sounded the alarm on anthropogenic climate change and drew attention from the media, policymakers, and the wider public. It comes as no surprise, then, that the CEI received substantial funding from ExxonMobil and the American Petroleum Institute (Shakir 2006). Certainly, the experts and public relation specialists paid for by the CEI didn't adhere to the standards of good scientific practice. They weren't interested in providing or scrutinizing scientific evidence. But should the denial of scientific facts be considered a crime? For Nye the answer might seem to be straightforward: "In these cases, for me, as a taxpayer and voter, the introduction of this extreme doubt about climate change is affecting my quality of life as a public citizen. (…) So I can see where people are very concerned about this, and they're pursuing criminal investigations" (Richardson 2016). And his concern is understandable, stemming from the potentially catastrophic consequences of inaction and the dangers of climate change for hundreds of millions of people.

The problem with such a depiction, however appealing it might be to climate scientists and proponents of a robust global climate change policy, is that it overlooks the complicated and problematic relations between expertise and decision-making, and science and politics. The situation certainly might seem unambiguous: Scientific experts, sounding the alarm on global climate change, are desperately trying to speak "truth to power", but, unfortunately, power seems unwilling or incapable of listening. The "truth" on anthropogenic climate change is drowned out by deliberate political misinterpretation of data and facts, alternative theories that never meet the standards of good scientific practice, and, perhaps worst of all, false statements, studies, and reports that resemble scientific research.

But this depiction presupposes a clear-cut conflict between objective bearers of true knowledge on the one side and a group of interest-driven distorters or knowledge on the other. In this picture, scientific facts become politicized if and when they transgress the boundaries between the aseptic laboratories of scientific research and enter the battlefield of political ideas, intrigues, and interests. But what if the science itself is contested? Is it always easy to separate healthy and rigorous scientific questioning from politically motivated distortion? How can one reliably discern truth from error? How can it be decided which scientific expertise to trust? And, even more importantly, how might one tell unintentional error form the intentional misinterpretation of scientific data?

Investigating how think tanks are involved in these processes of translating or distorting scientific findings is the starting point of this book.

These questions concerned the American public half a century ago, when Herman Kahn, a staff member of the RAND Corporation and later a founder of the conservative "Hudson Institute", published a comprehensive monograph titled *On Thermonuclear War*. (Kahn 1960) The book was remarkable at the time for including a detailed expert analysis on how to deviate from the doctrine of mutual assured destruction (MAD) in order to win a nuclear confrontation. The book was widely discussed and frequently criticized for bringing the possibilities of nuclear war closer. When Kahn died in 1983, *The New York Times* cited in his obituary a critical *Scientific American* editorial accusing *On Thermonuclear War* to be "a moral tract on mass murder: how to plan it, how to commit it, how to get away with it, how to justify it" (Treaster 1983).

Kahn's alarming analysis was widely condemned. According to some of his numerous critics, his advice could have threatened the lives of vast numbers of human beings. However, although Herman Kahn was never sued or publicly shamed for his writings, his controversial advice contributed to discrediting "megadeath intellectuals"[1] (Menand 2005). So, what's the difference between his proposals on the use of so-called doomsday devices, technologies, that is, that could destroy all human life on earth, and more recent warnings of the continued use of fossil fuel-based technology, which might bring equally unpleasant consequences? Was it that the use of nuclear weapons posed an *immediate* threat, that the horrors of thermonuclear war threatened the readers of Kahn rather than their children or grandchildren? Or was it the iconic image of the mushroom cloud that spurred emotional reaction to Kahn's cool, distanced, and scientific analysis? The risks of a nuclear war were arguably more tangible than the risks of a changing climate. Certainly, climate scientists have to overcome great obstacles when "sounding the alarm" on anthropogenic climate change. Is this why the "scandal" surrounding Bill Nye failed, in the end, to make major headlines and was soon displaced by other news stories? Distrusting the predictions of some scientists is, in the end, different from ignoring the obvious threats of nuclear weapons, or the proven dangers of smoking. Isn't it?

One obstacle for climate change policy proponents is that many believe climate change to be a problem that concerns people "someday in the future". While the devastating consequences of atomic bombs were entirely foreseeable, predicting and pinpointing the impact of climate change are far more difficult. Unable to "prove" that a single extreme weather event, such as Hurricane Katrina of 2005 or Hurricane Harvey of 2017, was caused by climate change, scientists have to turn to probabilities and models to explain the complex interplay of increasing water temperatures and atmospheric water vapour and pointing out that 'the strongest storms will become more powerful this century' (Hansen 2009: 253).

Nevertheless, scientists feel they have a moral obligation as a scientist to inform the wider public of the potential consequences. In 2009, for example, renowned climate scientists James Hansen published a book which aimed at telling "[t]he truth about the coming climate catastrophe and our last chance to save humanity". The book's somewhat sensational subtitle *Storms of My Grandchildren* encapsulates its urgent and deeply personal message. As Hansen explains: 'I did not want my grandchildren, someday in the future, to look back and say *Opa understood what was happening, but he did not make it clear*' (2009: XII).

His sentiment of ethical responsibility is one reason for the growing frustration of the international community of climate scientists, activists, and commentators such as Bill Nye, with climate sceptics and deniers. Yet it is no coincidence that the "scandal" took place in the US. As we will see in the course of this book, climate science is particularly contested in the US where political camps disagree sharply over climate politics. And while climate science is an international undertaking, driven mainly by an international community of climate scientists and international bodies like the Intergovernmental Panel on Climate Change, climate politics and consulting on the issue still take place at a predominantly national level.

In 2014, the year before the United Nations (UN) Climate Change Conference in Paris (COP 21), the comprehensive "The GLOBE Climate Legislation Study" found that despite a general trend towards more ambitious national climate politics, some industrialized countries (most notably, the US, Australia, and Japan) had taken a step back (e.g. by lowering their emission reduction targets) in order to stimulate short-term economic growth (Nachmany et al. 2014). Likewise, despite the attempts of international community of scientists to target a global audience, awareness, acceptance, and attitudes towards climate science differ considerably between countries. Climate change denial thrives particularly in the US.

This differentiation reveals how climate change is a highly political issue that requires the commitment of policymakers. The publication of scientific research simply isn't enough. We know that the Earth's climate is changing because scientists, such as James Hansen, have carefully gathered and analysed data. They have not only discovered changes in global climatic patterns but also attribute these changes to human activity, primarily the massive increase of carbon dioxide emissions from the burning of fossil fuels. Yet despite the lack of any real disputes within the community of climate scientists, the recipients of climate knowledge—policymakers, journalists, activists, and citizens—still doubt its validity. This situation might seem particularly puzzling since scientific research isn't challenged or contested by other scientists. Since we are said today to live in "knowledge societies" (Stehr 1994)—societies, that is, that are increasingly reliant on (scientific) knowledge—this apparent neglect of a scientific consensus needs further explanation.

It may well be, in fact, that the answer lies, at least partly within this depiction of modern societies. As Robert Proctor points out, actively spreading ignorance, raising doubt, and questioning scientific findings become a particularly valuable "strategic ploy" in social and political contexts used to looking for scientific expertise to plan and/or legitimize political action (Proctor 2008: 8–9).

Gone are the days when the *Machinery of Government Committee* (referred to as "Haldane Commission" after its chairman, the Viscount Haldane of Cloan) which was tasked with developing guiding principles for British research policy could state: 'It appears to us that adequate provision has not been made in the past for the organised acquisition of facts and information, and for the systematic application of thought as preliminary to the settlement of policy and its subsequent administration' (The Machinery of Government Committee 1918: 6). This Commission not only held policymakers accountable for acquiring the best information but also proposed far-reaching competences for researchers to decide on the public funding of scientific research. The context of science in society has changed drastically since then. Scientific expertise can still provide guidance, raise awareness, and offer solutions. But, as will be shown, expertise is also now regarded as a service, demanded and paid for by policymakers and interests groups who seek legitimization for their convictions and preferences.

This becomes particularly evident in climate politics. Controversies here arise not only between climate change deniers and scientists but between climate change deniers and climate change "believers". To believe in climate change isn't necessarily the same as to know or to understand the underlying science. It indicates that people are convinced that climate scientists 'got it right'.[2] But what convinces people that experts are right, that scientists actually do speak truth to power?

Communicating science is neither easy nor does it necessarily have to rest on the best arguments in order to get attention. Individual reputation, the prestige of a research facility or university, but also personal charisma and rhetorical skills can help determine the success of science communication. Moreover, it involves much more than simply *explaining* the complex implications of scientific discoveries to people outside the scientific community. Science communication must aim at *translating* scientific knowledge (e.g. climate knowledge) to audiences who might have their own (non-scientific) understanding, prefer specific interpretations, and have a political interest in climate research.

This book will focus on a particularly crucial and interesting type of research organization, purposefully built to communicate with policymakers and the public: think tanks. It will investigate their respective role in two selected countries, Germany and the US, and attempt to link to differences in national climate politics to distinct ways, opportunities, and constraints in the distribution of climate knowledge. Focusing on think tanks promises to augment our understanding of why climate knowledge is more generally unequally distributed and received. Moreover, investigating the roles of think tanks helps understanding how the translation of (abstract) findings of scientific research into politically "useful" knowledge is organized.

Investigating the role of think tanks in communicating climate science and assessing their impact on climate politics is especially promising since the rise and global spread of these organizations can be explained by an increasing demand for scientific advice. The number of think tanks is rapidly growing around the world. But, as the book shows, think tanks aren't necessarily an indicator for an increasing "scientization", let alone "rationalization" of the policy cycle. Think tanks can serve a different purpose. Not only do they provide decision-makers with scientific knowledge, they are also said to be instrumental in defending political interests against scientific knowledge claims. In contrast to classical lobby organizations, think tanks are thus accused of "hijacking" scientific authority: 'When the Georg C. Marshall Institute began to challenge the claims of the scientific on the ozone hole and global warming, they didn't create their own journal, but they did produce reports with the trappings of scientific argumentation – graphs, charts, references, and the like' write Naomi Oreskes and Eric Conway (2010: 244) outlining a strategy that is successfully pursued by a number of conservative, "climate change sceptics" in the US.

However, there are important reasons for why an investigation into think tanks is also particularly challenging. Not least because despite having attracted considerable scholarly attention, what is lacking from think tank research is an accepted definition of what exactly is the object of analysis (Weaver 1989: 563). International comparative studies, like the Global Go To Think Tank Index (https://www.gotothinktank.com) therefore use a wide definition to trace the "global spread" of these enigmatic organizations. This book aims at refining the prevailing characterization of think tanks. By investigating and comparing their respective role in the national climate politics of Germany and the US, it will consider the varieties at an organizational level. Although it focuses on climate politics, it can provide more widely applicable insights regarding the respective strategies, clients, and target audiences of think tanks.

In sum, the book intends to contribute to a better understanding of the role played by think tanks on climate politics in distinct political environments. The book is therefore divided into a theoretical section (Chaps. 2 and 3) and an empirical section that focuses specifically upon the impact of think tanks on climate politics (Chaps. 4, 5, and 6).

Structure of the Book

Climate change is scientific term describing the complex interplay of various variables affecting local weather, regional and seasonal weather patterns, and, finally, the Earth's climate. Anthropogenic climate change is a refinement of the scientific term indicating a major factor of the causation of global climate change. At the same time (and for this reason) anthropogenic climate change has become a political slogan, heavily contested not so much by scientific peers but by political parties, partisan experts, representatives of business interests, and, not least, think tanks.

Chapter 2 therefore discusses the complex relation of climate, knowledge, and politics. Climate change, if taken seriously, demands robust policymaking. This raises the important question of how exactly science can inform decision-making and how to safeguard science from being "politicized" and whether, indeed, this is actually possible. The chapter describes some of the principles and strategies that allow climate sceptics to beat climate scientists at their own game: Exploiting the differences between an everyday understanding and a scientific definition of "certainty" and "proof" allows partisan experts advocacy think tanks to cast doubt, influence public opinion, and provide policymaker with "reasoned arguments" against climate protection.

Chapter 3 outlines the conceptual framework for analysing think tanks, assessing the effectiveness of their respective strategies, and estimating their impact on the national climate politics in the United States and Germany. I argue that differences in the respective political systems have to be taken into account in order to explain the different roles and strategies of think tanks in the United States and Germany. Moreover, different "knowledge regimes" explain why think tanks differ at the organizational level and why, for instance, partisan advocacy think tanks, arguably the most important type in the US, struggle to getting influence in Germany.

Chapter 4 focuses on think tanks in the United States. The chapter provides an historical overview of the evolution of think tank in the United States with a special focus on how think tanks became instrumental in serving special political interests, including climate politics.

Chapter 5 maps the think tank landscape in Germany highlighting structural and historical particularities that help explain why academic research institutions dominate while German think tanks in general stay more in the background.

Finally, Chap. 6 systematically compares these two cases, analysing the way in which think tanks "fit in" their respective institutional and political environment and how they shape public debate and political decision-making on climate politics. The chapter shows that the respective knowledge regime influences opportunities for think tanks to accumulate "social capital" and in consequence affects the strategies that are available to them.

Ultimately, the study shows that understanding the significance and the consequences of the "global spread of think tanks" (McGann 2016) requires a comparative approach that doesn't focus only on the organizational qualities of think tanks but takes into account the complex institutional context and distinct "political cultures".

Notes

1. The term "megadeath intellectuals" was coined in the 1960s and refers to academics who (working at prestigious universities or think tanks such as the RAND Corporation) were actively involved in producing applicable knowledge to support policies of nuclear deterrence (see Raskin 1963).
2. However, accepting climate science (for whatever reasons) does not exempt one from making political, that is, normative decisions (see Machin 2013: 11).

References

Hansen, James. 2009. *Storms of My Grandchildren. The Truth About the Coming Climate Catastrophe and Our Last Chance to Save Humanity*. London/Berlin/New York: Bloomsbury.

Kahn, Herman. 1960. *On Thermonuclear War*. Princeton: Princeton University Press.

Machin, Amanda. 2013. *Negotiating Climate Change. Radical Democracy and the Illusion of Consensus*. London/New York: Zed Books.

McGann, James. 2016. *2015 The Global Go To Think Tank Index Report*. Think Tanks and Civil Societies Program, September 2.

Menand, Louis. 2005. Fat Man. *The New Yorker*, June 27. https://www.newyorker.com/magazine/2005/06/27/fat-man. Accessed 16 Nov 2017.

Nachmany, Michal, Samuel Fankhauser, Terry Townshend, Murray Collins, Tucker Landesman, Adam Matthews, Carolina Pavese, Katharina Rietig, Philip Schleifer, and Joana Setzer. 2014. *The GLOBE Climate Legislation Study: A Review of Climate Change Legislation in 66 Countries: Fourth Edition*. London: GLOBE International and Grantham Research Institute, LSE.

Oreskes, Naomi, and Erik, Conway. 2010. *Merchants of Doubt: How a Handful of Scientists Obscured the Truth on Issues from Tobacco Smoke to Global Warming*. New York/Berlin/London: Bloomsbury.

Proctor, Robert N. 2008. Agnotology: A Missing Term to Describe the Cultural Production of Ignorance (and Its Study). In *Agnotology. The Making & Unmaking of Ignorance*, ed. Robert N. Proctor and Londa Schiebinger, 1–36. Stanford: Stanford University Press.

———. 2012. The History of the Discovery of the Cigarette-Lung Cancer Link: Evidentiary Traditions, Corporate Denial, Global Toll. *Tobacco Control* 21: 87–91.

Raskin, Marcus G. 1963. Megadeath Intellectuals. *The New York Review of Books*, November 14.

Richardson, Valerie. 2016. Bill Nye, the Science Guy, Is Open to Criminal Charges and Jail Time for Climate Change Dissenters. *The Washington Times*, April 14.

Shakir, Faiz. 2006. Big Oil Launches Attack on Al Gore. *ThinkProgress*, May 17. https://thinkprogress.org/big-oil-launches-attack-on-al-gore-6f95e972303c. Accessed 10 July 2017.

Stehr, Nico. 1994. *Knowledge Societies*. London: SAGE.

The Machinery of Government Committee. 1918. *Report of the Machinery of Government Committee*. London: His Majesties Stationery Office.

Treaster, Jospeh B. 1983. Herman Kahn Dies: Futurist and Thinker on Nuclear Strategy. *The New York Times*, Obituary, July 8.

Weaver, Kent. 1989. The Changing World of Think Tanks. *PS: Political Science and Politics* 22: 563–578.

CHAPTER 2

Knowledge and Climate

Introduction

On 1 June 2017, President of the United States Donald J. Trump announced that the US would pull out of the 2015 Paris Agreement, arguably the most ambitious, and certainly the most applauded, effort by the international community to deal with the problem of global climate change. Donald Trump's attempts to cast the decision as a "reassertion of America's sovereignty" was immediately criticized at home and abroad.

Trump gave political and economic reasons for his decision. According to the White House, the Paris Agreement would undermine US competitiveness and would cost the US economy jobs while at the same time diverting precious taxpayers' money to an illegitimate international climate fund. The Obama Administration was accused of having negotiated a "bad deal" which would, in any case, have only a negligible impact on climate change (cf. https://www.whitehouse.gov/blog/2017/06/01/president-donald-j-trump-announces-us-withdrawal-paris-climate-accord).

It is worth noting, however, that the announcement failed to address the potential risks and dangers of climate change. In fact, it hardly mentioned climate change at all. In contrast, the then newly elected French President Emmanuel Macron stated that Trump had 'committed an error for the interests of his country, his people and a mistake for the future of our planet' (Watts and Connolly 2017). He further re-emphasized his

A. Ruser, *Climate Politics and the Impact of Think Tanks*,
https://doi.org/10.1007/978-3-319-75750-6_2

country's commitment to the Paris Agreement adding 'Don't be mistaken on climate: there is no plan B because there is no planet B' (ibid.).

While Trump's controversial statements tend to get the most attention, the French president's notion is claimed to actually merit more close analysis. Macron expresses an opinion which is widespread among climate scientists, environmentalists, and pro-climate politicians, according to which the challenge of climate change indicates "exceptional circumstances" (Stehr, Ruser 2018) and involves "exceptional dangers" (ibid.) and thus should somehow evade the rules and restrictions of "ordinary" political problems.

But can we speak of "politics" or "policies" when there's no plan B? For, the existence of a plan B implies the availability of political alternatives, which, in turn, is a necessary prerequisite for democratic decision-making.

In the classic definitions in the social sciences, *politics* understood as the activities of government and political bodies and *policies* which include a more specific set of ideas, concepts, or plans involve an element of active choice. Democracy depends on choice, that is, on the existence of alternative plans B, C, or D. It is important to keep this in mind for debates on climate change and climate politics sometimes emphasize a more technocratic vision of policymakers accepting the "truth" of climate change act according to and in line with the advice given to them by scientific experts.

Trump's announced withdrawal from the Paris Agreement is certainly just the latest episode in the difficult and sometimes frustrating history of international climate politics. The most important lesson that can be learned from this is that climate science and climate politics are inextricably intertwined, and dealing with climate change is essentially a political issue.

Before Trump's decision to pull out of the Paris Agreement was even officially confirmed, Jerry Brown, governor of California and outspoken proponent of action of climate change, criticized Trump's views, claiming that we 'can't fight reality' (Chaitin 2017). Brown's comment includes a strong yet fairly common accusation that dissenting from climate science is not simply a political position but rather indicates that the dissenter is actually out touch with "reality".[1]

From this perspective, withdrawing from the Paris Agreement and denying the existence of climate change therefore isn't just wrong, or economically and politically disadvantageous, but is actually insane!

Donald Trump's withdrawal from the Paris Agreement is therefore much more than just the latest chapter in global climate politics. It exemplifies the complicated relation between climate science and climate politics, between belief and denial and the special role of knowledge and trust. Furthermore, the readiness to trust scientific analysis and to follow recommendations issued by scientist based on their research might be directly linked to the salience of a problem. Rolf Lidskog, for instance, argues 'that when risks are imperceivable people may maintain their ontological security by ignoring experts' recommendations and not changing their routines for the organizing and monitoring of everyday life' (Lidskog 1996: 33). This leaves us with the important issue of whether climate change is salient and when and how it might become so?

But even if not, debate and dispute on climate change and climate politics involve the question of whether addressing climate change should be a political question. If climate change is about natural processes like the complex interplay of atmospheric gases, surface albedo, and the thermal expansion of the oceans, can there be room for promises, agreements, backroom deals, or, in short, politics? Renowned climate scientists and former director of NASA's Goddard Space Institute for Space Studies James Hansen, for instance, seems to see the room for political manoeuvring significantly constrained and calls for an acceptance of global climate change by urging lawmakers to comply with this new reality:

> Politicians think that if matters look difficult, compromise is a good approach. Unfortunately, nature and the laws of physics cannot compromise — they are what they are. (Hansen 2010)

How then should the relation between climate science and climate politics be conceptualized? Should we think of the issue of a scientific "fact", which compels political action? Or should we think of climate change as a political problem to be solved by policymakers who may or may not believe in climate change? Is dissenting on climate change a legitimate political position or indicating an alarming denial of reality?

To fully understand the dispute over the "reality" of climate change, to comprehend the political strategies and finally the role of scientific research institutes and (political) think tanks, we have to begin with a discussion of the respective roles "knowledge" can play in public discourses and political decision-making.

This chapter therefore considers the connection between scientific knowledge and climate politics. It focuses particularly on the way in which knowledge of climate change is acquired and on the question of what it means to “trust science” and how this trust arises. It suggests that a great deal hinges upon what is understood as qualifying as “knowledge” and how the differences in the scientific and lay conceptions of this word allow a deliberate undermining of trust in science.

The theoretical discussion in this chapter will form the basis for the more thorough analysis of the role and the impact of (environmental) think tanks as switchboard organizations tasked not only with disseminating knowledge but also to provide support for political and normative positions. It will be demonstrated how think tanks manage to translate, disseminate, and, at times, distort climate knowledge and why they are important for formulating national climate policies and politics.

Climate Knowledge, Climate Belief, and Trust in Climate Science

How do we know that the climate is changing? How can we be sure that climate change isn’t a clever story made up by a bunch of egotistic climate scientists desperate to get attention and lucrative research grants? This is the suspicion recently articulated by Lamar Smith, Republican Chair of the United States Congress ‘House Science, Space and Technology Committee’ (cf. Cousins 2017). The claim is symptomatic of an apparently widespread distrust of climate change scientists and policy.

Intriguingly disputes about climate change are said to be raging between climate change *believers* and climate change *deniers* (cf. Waldholz 2017). What does this tell us? Should climate change and climate politics not hinge upon *knowing* rather than believing? After all, global, human-made climate change is among the most urgently researched topics today; the joint research agenda of the Intergovernmental Panel on Climate Change (IPCC) alone involves thousands of scientists from around the world.

How can we *know* that assessments such as that made by the IPCC—‘[i]t is *extremely likely* that human influence has been the dominant cause of the observed warming since the mid-20th century’ (IPCC Summary for Policymakers 2013: 17)—reflect the best available knowledge? And, perhaps even more importantly, how can we know that human activity is the main cause of climate change? Can (alleged) laypersons ever hope to understand the complex science behind the "facts" of global climate change? Or are policymakers and their voters alike depended on believing in science and plan "rational" or "feasible" solutions accordingly?

Climate change is a particularly difficult phenomenon to know. This is because climate change can neither be experienced directly nor can it be separated from social, economic, and even cultural aspects. What does it mean to say that climate change cannot be experienced directly? It is useful here to turn to the definition of "climate" provided by the IPCC:

> Climate in a narrow sense is usually defined as the "average weather," or more rigorously, as the statistical description in terms of the mean and variability of relevant quantities over a period of time ranging from months to thousands of years. The classical period is 3 decades, as defined by the World Meteorological Organization (WMO). These quantities are most often surface variables such as temperature, precipitation, and wind. Climate in a wider sense is the state, including a statistical description, of the climate system. (http://www.ipcc.ch/ipccreports/tar/wg2/index.php?idp=689)

According to the IPCC definition, then, "climate" is a scientific construction, used to describe statistically significant average conditions that don't actually exist in reality. Observable data, for example, a specific weather event, the amount of rainfall in a given area and period of time, wind speed, or the daily maximum temperature, should be thought of as small components which, when fed in sophisticated scientific models, allow a scientific description of the Earth's climate—or, for that matter, changes in the average climatic conditions.

Climate change is a scientific label used to describe the systematic interconnection of events and patterns in the world's climatic system too.[2] Accordingly, the IPCC gives a rather detailed definition of climate change specifying not only the data that is observed and included but also the means by which these observations are made:

> *Climate change in IPCC usage refers to a change in the state of the climate that can be identified (e.g. using statistical tests) by changes in the mean and/or the variability of its properties, and that persists for an extended period, typically decades or longer. It refers to any change in climate over time, whether due to natural variability or as a result of human activity. This usage differs from that in the United Nations Framework Convention on Climate Change (UNFCCC), where climate change refers to a change of climate that is*

attributed directly or indirectly to human activity that alters the composition of the global atmosphere and that is in addition to natural climate variability observed over comparable time periods. (IPCC 2007)

The definition given by the IPCC draws an important distinction between a more comprehensive, scientific understanding of climate change as including both, change stemming from human activity and natural causes alike, and the more narrow definition of human-made or "anthropogenic" climate change used by the UNFCCC. Moreover, it reflects the methodological approach as well as the empirical obstacles climate science has had to face for decades.

Observing the climate is difficult and costly. In the "technical memorandum 86152" from 1984, NASA scientist Lloyd A. Treinish outlines the features and advantages of the newly established 'General Scientific Information System to support the study of climate related data', most notably the 'Pilot Climate Data System (PCDS)' (Treinish 1984: 1). The technical memorandum is interesting since it showcases the various practical difficulties modern climate science has had to overcome in order to detect and secure changes in the world's climatic system.

Climate research depends on collecting, preparing, and analysing vast quantities of data that have to be collected on a global scale and over an extended period of time. The PCDS as described in the memorandum promised to provide a solution to these challenges by using (then modern) computer systems (ibid.: 4), satellite surveillance, and modern database management software to deal with a vast quantity of measurement data.

The PCDS also gives an insight in the daily practice of climate research. While contemporary climate scientists no longer rely on 'a computer system based upon a Digital Equipment Corporation (DEC) VAX-11/780' (ibid.: 4), computer modelling and analysing large quantities of observational data are still integral parts of climate research. Moreover, and perhaps even more importantly, climate change research is relying heavily on data (time series) of past events in order to understand the complex interplay of the global climatic system and ultimately predict (likely) future developments.

In a classical paper from 1957, pioneers of modern climate science Roger Revelle and Hans E. Suess attempted to estimate the concentration of atmospheric CO_2 by comparing scientific calculation based on nineteenth-century measurement with more recent, that is, in this case, early- to mid-twentieth-century data (Revelle and Suess 1957: 18–19). The scientific article is a good example for the difficulties and uncertainties stemming from the complexity of the Earth's climatic system (e.g. uncertainty about the amount of CO_2 that is absorbed and stored in the world oceans, ibid.: 19–20) and its impact of estimating future developments. Instead of deriving or predicting a definite future atmospheric concentration of carbon dioxide, Revelle and Suess refer to a technique that is still the method of choice for climate scientists: They outline a set of future *scenarios* (Revelle and Suess 1957: 25).

Climate researchers have to rely on complex methods of data analysis and equally complex techniques of scenario development because, as said above, climate change cannot be experienced directly. It is because of the complexity of the climatic system that climate scientists are often "building worlds" by making 'use of computerized climate models as tools for producing knowledge of the earth's climate' (Miller and Edwards 2001: 16). But what kind of climate knowledge can be derived from computerized climate models?

Within the scientific community the availability of efficient means of data processing clearly indicates progress. Powerful computers allow for building ever more sophisticated climate models. This is very important since '[t]he data record of climate (…) is sparse, incomplete, and poorly fitted to modelling grids. Computer models are used to integrate, filter, smooth, and interpolate these data, building uniform, consistent global data sets that now form the basis of knowledge about climatic change over the last century' (ibid.: 17).

Basing climate knowledge on models and simulations highlights the tentative nature of scientific knowledge. The provisional nature of knowledge derived from climate knowledge is not regarded a problem within the community of climate scientists. Since scientific findings are inevitably preliminary and prone to falsification by further research, climate scientists are more concerned with *improving* their models than with arriving at "definite" answers. Accordingly, a chapter for the Fourth Assessment Report of the Intergovernmental Panel on Climate Change on "Climate Models and Their Evaluation" focuses on the climate models' capacity for "projecting future climate change" (Randall et al. 2007: 590).

> Climate models are based on well-established physical principles and have been demonstrated to reproduce observed features of recent climate (…) and past climate changes (…). There is considerable confidence that Atmosphere-Ocean General Circulation Models (AOGCMs) provide credible quantitative estimates of future climate change, particularly at continental and larger scales. Confidence in these estimates is higher for some climate variables (e.g., temperature) than for others (e.g., precipitation). (Randall et al. 2007: 591)

It is important to stress that different levels of confidence for some variables are not problematic for the production of scientific knowledge about climate change. Rather the contrary, lack of knowledge is understood to be part of the process. To understand the complex interplay of the various variables that add up to the processes that are subsumed under the category of "climate change" requires the steady improvement and refinement of climate models:

> The large-scale patterns of seasonal variation in several important atmospheric fields are now better simulated by AOGCMs than they were at the time of the TAR. Notably, errors in simulating the monthly mean, global distribution of precipitation, sea level pressure and surface air temperature have all decreased. In some models, simulation of marine low- level clouds, which are important for correctly simulating sea surface temperature and cloud feedback in a changing climate, has also improved. Nevertheless, important deficiencies remain in the simulation of clouds and tropical precipitation (with their important regional and global impacts). (Randall et al. 2007: 592)

As stated above, climate models are crucial for *understanding* the complex processes and mechanisms of a changing climatic system. Relying on "models", "scenarios", and "simulations" therefore does not indicate a deficit. Unlike their use in everyday language, complex simulation and computer modelling are epitomizing progress in the context of climate science:

> Only through simulation can you systematically and repeatedly test variations in the "forcings" (the variables that control the climate system). Even more important, only through modelling can you create a control—a simulated Earth with pre-industrial levels of greenhouse gases, or without the chlorofluorocarbons that erode the ozone layer, or without aerosols from fossil fuel and agricultural waste combustion—against which to analyse what is happening on the real Earth. (Edwards 2010: 140)

As Paul Edwards (2010) has shown in his comprehensive, historical analysis *A Vast Machine: Computer Models, Climate Data, and the Politics of Global Warming*, it was the availability of sophisticated digital computing devices which made truly scientific observation of the Earth's climate possible. Far from indicating a deficit, computer models have helped researchers *overcome* a fundamental deficit, for they allowed controlled scientific experiments to be conducted. Before the advent of powerful computers, experimenting was all but impossible in the "geophysical" or "climate" sciences. One reason is the complexity of the object of research:

> The system you are dealing with is just too large and too complex. You can isolate some things, such as the radiative properties of gases, in a laboratory, but to understand how those things affect the climate you need to know how they interact with everything else. (Edwards 2010: 139)

A second, even more important, obstacle was the impossibility to guarantee "controlled conditions". Observing the global climatic conditions would require global laboratory conditions, that is, the means to compare the data obtained by experiment on the global scale. This, of course, is impossible, since:

> There is no "control Earth" that you can hold constant while twisting the dials on a different, experimental Earth, changing carbon dioxide or aerosols or solar input to find out how they interact, or which one affects the climate most, or how much difference a change in one variable might make. (ibid.: 140)

Computers enabled researchers to "building worlds" (Miller and Edwards 2001: 17), that is, to construct a "virtual" control Earth which allowed for all the comparison, repetition, and selected tests of variables necessary for experimental research:

> Simulation modeling opened up a way out (...). Only through simulation can you systematically and repeatedly test variations in the "forcings" (the variables that control the climate system). Even more important, only through modeling can you create a control—a simulated Earth with pre-industrial levels of greenhouse gases, or without the chlo- rofluorocarbons that erode the ozone layer, or without aerosols from fossil fuel and agricultural waste combustion—against which to analyze what is happening on the real Earth. (ibid.)

The advent of modern, more powerful means to process large quantities of data thus clearly indicated decisive advancement for climate science. Yet at the same time it contributed to setting climate science further apart from one's everyday experience: Climate change research became more scientifically sound and opaque at the same time. For this reason, climate science is characterized by a *double inaccessibility*:

Its objects of research, climate and climate change cannot be experienced directly but must be "read off" from statistical or observational data. At the same time, the scientific methods required to do this "reading off" grow ever more complex making it virtually impossible for anybody outside the community of climate researches to judge the quality of the findings.

This double inaccessibility raises the problem of "trust in science". In fact, the problem is even more complicated because it does not only involve the question of *how* trust in science can be established. We also have to ask *why* one should trust scientific knowledge claims.

The everyday understanding of science seemingly correspond closely with the philosophical theory of "scientific realism", which starts from the conviction that (natural) science 'provides descriptive, accurate information about the physical reality' (Rescher 1987, cited after Ruser 2018: 770), thus allowing scientists to 'tell it like it really is'. Such a "naïve" perspective assumes that science has authority because and to the degree that it can tell "for sure" how it really is. According to H.M. Collins, this authority stems from the fact that people can become "virtual witnesses" of the research process and can therefore 'see the validity of the procedures and findings' (Collins 1992: 160) In turn, trusting science might be particularly demanding, when the procedures are hard to observe and findings difficult to understand.

As discussed above, climate science deals with a highly complex, ultimately inaccessible object of study: the Earth's climate. Moreover, as we have shown, the methods and procedures employed to grasp the "reality" of climate change are equally complex and difficult to understand for anyone not specialized in this particular field of research. In consequence, accepting the findings of climate science requires a trust in the scientific method and the integrity of the climate scientist that, for the vast majority, is *not* rooted in personal observation or "witnessing" of the research process nor an individual understanding of the research findings. The problem is then how to assess the quality of scientific research and how to tell scientific knowledge claims form unscientific ones? Some try to solve this problem by referring to social criteria, for example, declaring that valid scientific knowledge 'is what is accepted by the scientific community' (Zimmerman 1995: 21). Unfortunately, such definitions simply beg the question; when, why, and how should one trust scientists to determine their findings are trustworthy.

Fortunately, philosophy of science does provide us with some criteria to assess the quality of scientific research and to tell scientific knowledge claims from non- or pseudoscientific claims. For instance, in order to be regarded as scientific, a knowledge claim 'has to be explanatory by reference to natural law' and be 'testable against the empirical world', and, most importantly, "it is falsifiable" (ibid.). The last criterion logically implies that scientific knowledge claims can be (and are) preliminary and should be labelled as "not-disproved" rather than "true".

This essential feature of science has major implications for the social role of science and, apparently, for scientific authority:

> "The tentativeness of scientific conclusions is what makes being a scientist and studying about science so much fun. It is also what makes scientist and a student of science so difficult. Because our understanding of the natural world is always open to improvement and modification, *we cannot rely with certainty on scientific explanations*" [emphasis added: AR] (Zimmerman 1995: 21). That we cannot rely on scientific explanation "with certainty" doesn't mean, that we cannot rely on them "at all". In fact, people rely on scientific explanation without definitive certainty all the time. There is a subtle difference in the scientific use of the term "certain" and the everyday understanding of certainty. While the everyday the use implies that an explanation, in order to be "certain" must be "the final word" (Zimmerman 1995: 21), scientists accept that new data, new theories and new models lead to new interpretations and new uncertainties. Or as Norman Miller puts it: "[S]cientists are not only comfortable with uncertainty, but literally build it into their thinking." (2009: 133)

It is important to note that this problem is *not* a particularity of climate science. Certain and trusted scientific knowledge has been disproven before, and at times with rather dramatic consequences.[3] However, since climate research is becoming increasingly relevant in the political realm, there is more at stake.

Indeed, trusting climate science is sometimes portrayed as an essential prerequisite for preventing dramatic consequences. It seems that what is really at stake is not the academic matter of the epistemology of climate change but rather the lives, health, and wealth of millions of people, today and in the future. As Lidskog has pointed out one would not only put one's faith in the scientist's ability to come up with suitable solutions but also that the specific risk actually exists. For

> it is a distinctive feature of present-day threats to the environment that they are increasingly diffuse (difficult to delimit in space and time) and hard to grasp. In many respects, they are invisible to the lay person's perception and beyond the lay person's range of experience. In certain cases present threats are of such a "delayed action" type as to involve possible consequences only for future generations. (Lidskog 1996: 34)

Can science contribute to people accepting and, in consequence, reacting to invisible risks? The problem is whether people *trust* scientific expertise. As again Lidskog points out, science in general and technological developments driven by scientific progress 'have provided man with the opportunity to protect himself and control risk' (ibid.: 37). However, as Lidskog continues, scientific expertise and authority isn't "automatically" translated into trust. 'People's perception of and reactions to risks have to be understood as phenomena largely influenced by their contexts, as created by a complex of general and specifically local circumstances' (ibid.)

'[W]hy, when the scientific evidence for global warming is unequivocal, does only half the public accept this evidence?' asks James Powell in the preface of his 2011 *The Inquisition of Climate Science* before he continues to list the name of prominent media representatives and policymakers in the US who question the integrity of climate science and its protagonists (ibid.: 1–2).

These public challenges failed to resonate within the community of climate scientists, and yet the challenges made by those sceptical of the reality of climate change remain prominent (Powell 2011: 12)?

If, as we have seen, building trust in climate science is relatively demanding, then undermining trust in climate science should be relatively easy. This is precisely the aim of climate change sceptics who use (or rather abuse) the peculiar, sophisticated aspects of philosophy of science delineated above to attack and discredit climate science.

The philosophical and methodological principles of science provide sceptics with (legitimate) means to criticize mainstream science, point towards uncertainties, and ask for further and more through research. Ironically perhaps, if sceptics limit themselves to challenging scientific findings on scientific grounds, then they simply pay an invaluable service to (climate) science by provoking further inquiry; "trust" in scientific research should not be confused with blind trust in scientific authority.[4]

However, if the attacks on climate science are motivated not by an interest in the science itself and its reliability but by political and economic interests, then climate science becomes a political plaything.

The Politicization of Climate Science

Even when some principles of the philosophy of science and some particularities of climate science provide sceptics and deniers the means to challenge its findings, it has to be asked *why* one would challenge scientific authority especially when it comes to climate science? For, other fields of research might be equally inaccessible. It could be asked whether climate scepticism and climate denial indicates a general rejection of scientific authority (which would include medical research or economic expertise) or whether the implications of climate research are posing a threat to special interests or challenging normative convictions. In short, it has to be asked what makes climate science (and climate scientists) a target for dissenting views.

Parallel to application of ever more complex and sophisticated methods, climate science became also more politically relevant and, in the wake of this development, itself increasingly politicized:

> Much of the early communication was relatively narrowly focused on scientific findings and synthesis reports (such as those published periodically by the Intergovernmental Panel on Climate Change, IPCC), sometimes occasioned by particularly severe extreme events, sometimes by high-level conferences or policy meetings. (Moser 2010: 32)

Likewise, scientific debate among climate scientists ceased to be purely academic. Early debates focused on how to improve the quality of methods and how to validate findings derived from the available evidence:

> Scientists have disputed complex issues of climate change detection, attribution (natural vs. human causes), and consequences, from how to validate climate models to whether rapid global warming might cause melting of the West Antarctic ice sheet or a shift in the North Atlantic current. They have sought to detect climatic changes, first in a warming "signal" in long-term, globally averaged temperatures and now, as well, in the statistical "fingerprint" of subtler shifts in climatic patterns. (Miller and Edwards 2001: 2)

However, with the consequences of a changing climate becoming an ever more prominent political issue, the pressure on climate scientist to quantify, that is, to number the degree of certainty and the potential consequences also grew:

> Growing demands for policy action have prompted increasingly complex and sophisticated attempts to quantify the potential damage form climatic changes and the potential costs of policies to prevent them. (ibid.)

It is understandable that climate scientists and policymakers alike want to "know for sure" whether the climate is changing and expect scientist to estimate the future risk of a changing climate. However, by complying to these expectations, climate science became even more vulnerable to attacks from outside the academic world. Studies and reports (most notably the IPCC Synthesis Reports, see below) were increasingly read by people from outside the scientific community and without the formal education required to catch the full complexity of climate research. The need to cater to the demand for comprehensible, easy-to-communicate information convinced climate scientist to provide "summaries for policymakers", estimate degrees of certainty, and apply labels of likelihood for specific consequence of climate change (cf. Machin and Ruser forthcoming). While these measures have certainly contributed to informing policymaking and the public, it hasn't necessarily increased the "trust" in climate science—for the need to communicating complex findings inevitably involves simplification. Emblematic numbers, for example, the 2C or 1.5C target(s), respectively (which tell dangerous from less dangerous climate change[5]), and most importantly the level of certainty that climate change is in fact human made are essential for *political* debates on climate change but at the same time allow critics, sceptics, and deniers to point at their arbitrariness and their political use:

> When the IPCC issued that statement that it 'is now 95% certain that humans are the main cause of current global warming' (IPCC 2014: v) it wasn't providing an "accurate" estimate but the results of a collation based on this scale. The important question, then, the one that can become decisive in disputes between climate change campaigners and climate change deniers, is whether the numerical expression is a useful and inevitable simplification or rather dubious a political move. (Machin and Ruser forthcoming)

The very fact that scientists are accepting these kinds of simplification should be regarded as an attempt to bridge the gap between scientists on the one hand and lawmakers on the other, a gap that is stemming from the two groups belonging to different "cultures":

> Lawmakers are faced with problems for which the public demands prompt solutions and cannot wait for definitive data, whereas science is patient and tentative. (...) Whereas differences of opinion among scientists are simply part of their everyday world, lawmakers see such conflicts as not only complicating their decisions, bit engendering public distrust, thus making anything they do suspect

writes again Miller (2009: 133), thus highlighting the delicate role of public opinion on any attempt to base political decision on scientific knowledge.

Climate Knowledge and Public Opinion

With climate scientist warning of dangerous, human-made climate change intensifying, climate politics gained importance in the national and the international level.

Since the early 1990s, a series of climate summits were held to establish and expand a global agenda with the Kyoto Protocol of 1997 being the most prominent and the Paris Agreement of 2015 the most recent attempt to define a binding set of rules for global climate politics, most notably emission reduction targets.

Yet, while forging an international alliance to address climate change has been quite successful, national politics and, equally important, public opinion in climate change are still diverging between countries:

> In 2009 in the *Special Eurobarometer Report 313* "European's Attitudes towards Climate Change" the European Commission found that while Europeans are in general believe climate change to be a problem younger, highly educated, left-leaning Europeans are particularly likely to consider climate change a major challenge of the future (Eurobarometer Special Report 313 2009: 14). There are considerable differences between the member states of the European Union when it comes to assess the urgency of the issue. While 94 per cent of respondents in Greece and 92 per cent in Cyprus respectively consider climate change a "very serious problem" only 49 per cent of respondents in Estonia and 51 per cent in the United Kingdom share this opinion. On average about two thirds (67 per cent) of citizens in the European Union are very concerned about climate change.

But what do these mere numbers indicate? How can these relative levels of concern be explained? The authors of the Eurobarometer Report 313

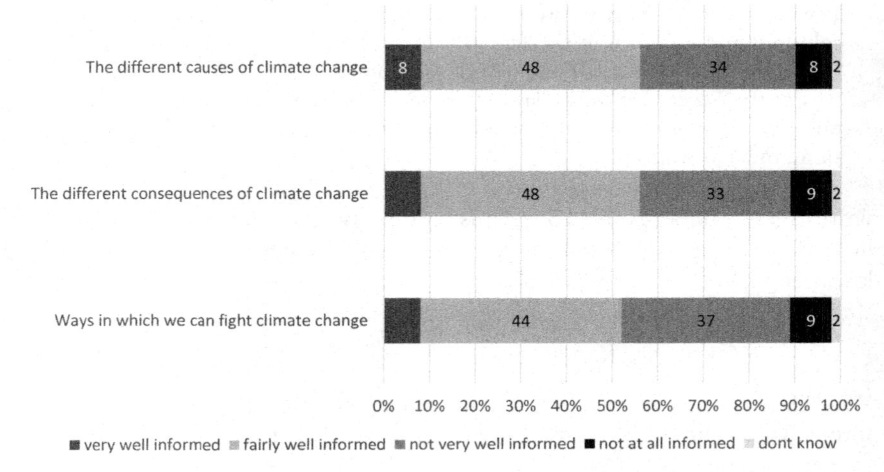

Fig. 2.1 European's attitudes towards climate change. (Source: Special Eurobarometer Report 313)

themselves point towards the close relation between the relative awareness of the issues and media coverage of climate-related topics.

Asking for the subjective level of information on the causes and consequences of anthropogenic climate change and the options of climate politics, the survey found that on average more than half of the respondents believe to be "very well" or "fairly well informed" with only eight to nine per cent feel not informed at all. The numbers are almost similar regardless of whether the question of *causes*, *consequences*, or different options for *fighting* climate change are addressed (Fig. 2.1).

The numbers show that Europeans, on average, feel well informed about climate change. But (how) does this level of information translate into a higher level of awareness? A representative survey on "Global Concern on Climate Change" conducted by the PEW Research Center in 2015 found considerable differences in public concern on climate change (Fig. 2.2).

Yet do, for instance, the numbers on European countries indicate a high level of *knowledge about* climate change or rather a widespread *belief in* a changing climate? Can, for example, national differences be attributed to an unequal distribution of knowledge about the "fact" of climate science, or do they just prove that disbelief in climate science is considerably higher in some countries? And, most importantly, how can these differences be explained?

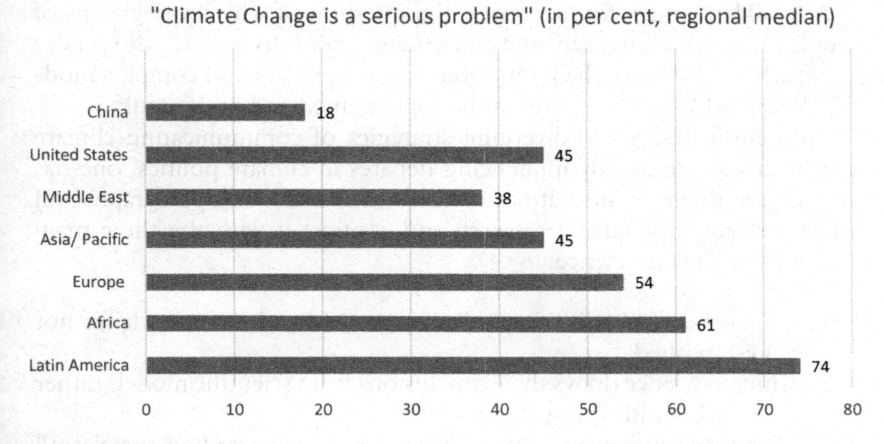

Fig. 2.2 Global Concern on Climate Change. (Source: Stokes et al. (2015). "Global Concern about Climate Change, Broad Support for Limiting Emissions". PEW Research Center)

These questions are particularly urgent since climate research is crucial, not only for offering suitable *solutions* but for identifying ("discovering") the problem in the first place. As we have seen above, unlike an extreme weather event like a flood, a storm, or a drought which might compel immediate political action because its devastating effects appear right before everyone's eyes, global climate change becomes "visible" in complex computer simulations, scientific graphs, and complex, even opaque data.

So why do some people believe in what they cannot see, while others don't?

Obviously, comparing the outcomes of international opinion polls isn't sufficient to answering this question. Other explanations (e.g. one's political conviction and/or party affiliation) have to be considered to describe a "typical climate sceptic" (see Chaps. 4, 5, and 6). Unfortunately, even if we would find that the typical climate sceptic in the United States is a white, protestant man, who lives in Middle West and votes for the Republican Party, we still wouldn't know what exactly makes him reject the findings of climate science? Why are he and his fellow sceptics not ready (yet) to accept scientific facts?

A feasible strategy for answering the question of why some groups of people believe in climate change and others don't is to analyse who is talking to whom. Who are the interpreters of opaque data and complex models? Who is addressed and how is the topic framed and presented?

To identify and assess diverging strategies of communicating climate science and subsequently influencing debates in climate politics, one has to compare them to the state of research. The following paragraphs will briefly present this state of research and contrast it with the three main criticisms of climate science:

1. Climate science is still a relatively new field and its main claims not yet established.
2. Climate science draws their conclusions from scientific models rather than real-world observations.
3. Climate scientists have a personal interest in "sounding the alarm" to attract grant money.

The Greenhouse Effect: A Nineteenth-Century Discovery

One of the most important misconceptions that is repeated in debates on global warming is that its main discoveries have been made only (fairly) recently. Climate science is portrayed as a young discipline, which, for some as we will see, needs to be developed much further before it should issue public reports or give far-reaching policy recommendations.

Such impressions of climate science as a discipline in a nascent state often stem from confusing climate *science* with climate *politics*. Climate change became a political issue especially in the last third of the twentieth century (cf. Jaspal and Nerlich 2012) eventually paving the way for the Brundtland Report issued in 1987[6] and the Earth Summit of 1992 in Rio de Janeiro.

Neither the report nor the conference came as an immediate response to a dramatic scientific discovery. The greenhouse effect in particular was known to the scientific community for more than a century when the Brundtland Report singled out an enhanced greenhouse effect to be an immediate threat to human life and well-being.

In fact, Irish scientists John Tyndall had designed a set of experiments leading to the discovery of the heat-absorbing qualities of certain gases in the atmosphere as early as 1859 (Hulme 2009: 121). While Tyndall was certainly a pioneer of modern climate science, he was not alone. Thirty years earlier the French mathematician Joseph Fourier had described the interchange between incoming solar radiation and outgoing "dark radiation" in the Earth's atmosphere much like a "glass covering a bowl" (Fourier [1824]1936 cited after Ruser 2012). Svante Arrhenius, (1896) eventual winner of the Nobel Prize for Chemistry, linked the increasing carbon dioxide concentration in the atmosphere to the industrialization that was taking place in North America in Europe and thus became the first to establish a link between human activity and the complex processes of atmospheric chemistry (cf. Ruser 2012).

Already in the mid of the twentieth century, Roger Revelle (1957) and Charles Keeling (1960) had established empirical evidence for an increase of CO_2 in the atmosphere due to human activity.

Moreover, in a classical paper from 1957, Roger Revelle and Hans Suess warned that the consequences of interfering with atmospheric chemistry could be far reaching and irreversible:

> Thus human beings are now carrying out a large scale geophysical experiment of a kind that could not have happened in the past nor be reproduced in the future. (Revelle and Suess 1957: 19)

Apparently, the greenhouse effect isn't a recent discovery (cf. Powell 2011: 36–37). Moreover, the impact of human activity most notably the burning of fossil fuel is a well- established, measurable finding since the late 1950s and early 1960s. Piling up evidence for human impact on the composition of the atmosphere climate scientists was increasingly heard by policymakers. In 1965 the US President Lyndon Johnson delivered a special message to Congress on the issue. In his speech, he expressed his conviction that '[t]his generation has altered the composition of the atmosphere on a global scale through (…) a steady increase in carbon dioxide from the burning of fossil fuels' (Johnson cited after Powell 2011: 43). In the mid-1960s human-made climate change was a scientifically proven fact.

A Model of Reality or a Lack of Real-World Observations?

As was already discussed above, climate science can rely only in part of real-world observations. To catch the complex interplay of the many variables that determine the climatic conditions on Earth and, even more importantly, to estimate the impact of changing greenhouse gas emissions on the future climate, scientists increasingly use complex computer models. From the perspective of scientists involved, this isn't a problem. The use of highly complex, yet still simplified, models to analyse and predict the interplay of an observable number of variables is consistent with the good scientific practice and also takes into account progress in the calculation capacities of modern computers. Of course, the use of models has some limitations. Climate scientists have to accept that for the time being, 'models can provide us only with a range of future possibilities' and that '[i]t will be a long time, if ever, before they can predict the effects of global warming, especially participation, on a regional scale' (Powell 2011: 142).

However, the use of models remains a popular target for critics of climate science. For instance, in the US, the conservative Heartland Institute criticized the use of models because, 'in spite of their sophistication, they remain merely models. They represent simulations of the real world, constrained by their ability to correctly capture and portray each of the important processes that affect climate' (Heartland Institute, Nongovernmental Panel on Climate Change 2013 "Global Climate Models and their limitations": https://www.heartland.org/publications-resources/publications/research--commentary-global-climate-models-and-their-limitations).

As in the case of "certainty", climate science has to deal with a subtle yet important difference between an everyday use and a scientific understanding of the term "model". Since you cannot use a model plane to fly from Berlin to New York, the model is obviously inferior to a *real* aircraft. While this is true when someone is planning her next business trip or vacation, for engineers who want to find out more about the aerodynamic qualities of a new aircraft design, a model would do just fine.

The same is true for climate science. Insights derived from running climate models might be wrong (but so might be conclusions based on real-world observation). Nevertheless, climate models might still be the best available means for scientists to find out more about the complex mechanisms that make the Earth's climate.

Moreover, as was outlined above, in climate science, the use of models is a mere necessity and even comes with some advantages (e.g. by allowing experimental testing of certain variables) rather than being inferior to real-world observation.

Climategate! How Can One Trust Science When Scientists Aren't Trustworthy?

On 28 November, *The Telegraph* reported the 'worst scientific scandal of our generation' demanding that '[o]ur hopelessly compromised scientific establishment cannot be allowed to get away with the Climategate white-wash'. What had happened? Obviously, there was (a) a scandal comparable in size and scope to the (in)famous Watergate affair that rocked the political establishment in the USA and (b) climate scientist seemingly attempted to covering the affair up.

What really had happened was this: In 2009 shortly before the Climate Summit in Copenhagen, a server of the Climate Research Unit (CRU) at the University of East Anglia (UK) was hacked, and files and emails had been stolen (see also Chap. 4). The files and mails were then leaked to climate sceptic websites. While most of the emails and files dealt with technical aspects of the scientific work, some were singled out to proof that climate scientists had something to hide and weren't reporting their full knowledge. As a matter of fact, some emails included complaints about some shortcomings of the available data:

> From: Kevin Trenberth (US National Center for Atmospheric Research). To: Michael Mann. Oct 12, 2009
>
> 'The fact is that we can't account for the lack of warming at the moment and it is a travesty that we can't... Our observing system is inadequate.'

In other emails climate researchers expressed their frustration that they had to answer to climate sceptics:

> From: Michael Mann. To: Phil Jones and Gabi Hegerl (University of Edinburgh). Date: Aug 10, 2004
>
> 'Phil and I are likely to have to respond to more crap criticisms from the idiots in the near future.'[7]

The leaked emails were far from uncovering the worst scientific scandal of a generation. Most people familiar with the research process wouldn't be surprised that researchers express their frustration about the quality of available data, thrash the work of some fellow researcher,[8] and complain about distraction from their work.

Nevertheless, the media attention caused an investigation of the behaviour of the scientists involved. Despite the fact that there was no credible evidence found for any wrongdoing (Powell 2011: 164–165) and worst scandal of a generation was soon to be forgotten, it reveals an important mechanism for casting doubt on the findings of climate research: emphasizing the social conditions of the research process.

Unfortunately, the scientific community do not consist of saints, and the scientific venture is as much a "quest for reality" (Tauber 1997) as it is a professional career. It goes without saying that scientists have an interest pushing their own careers or promoting "allies" in the quest for academic reputation. Also, scientific communities are deploying all sorts of instruments to safeguard scientific research form error guarantee adherence to the standards of good scientific practice.[9] Nevertheless, there has always been selfish behaviour or even fraud conducted by scientists (cf. Kohn 1986). Is it possible that individual climate scientist forges data to get personal gains? Absolutely yes. However, believing that an international community of climate scientists, consisting of many thousands of individual researchers, somehow conspire to get personal benefits is a different story.

What matters here is, as we will see in subsequent chapters, that challenging the integrity of climate scientist can be a suitable strategy to discredit climate science.

In all three cases, it is crucial how science, that is, mainly its findings and procedures but also the social and political context, is ***communicated***. The next paragraph therefore touches upon the role of the media in the formation of "public opinions" on scientific issues.

Disseminating Knowledge: The Role of the Media in Public Discourses

As we have seen in the previous paragraphs, science and scientific research is subject to many misunderstandings. The most important misrepresentation seems to be the 'belief that science is a collection of facts rather than an ongoing investigative process that permits us to gain insights and understanding into the way the world works' (Zimmerman 1995: 14). Since science isn't a collection of facts and because, in the case of climate science, its findings can have serious political consequences, the cardinal question is how to communicate the state of research. Reporting facts would be easy, but how about communicating a snapshot from an ongoing, vast, international research project?

Disseminating climate knowledge faces important challenges that stem from the complexity of the Earth's climatic system. Accordingly explaining the findings, underlying principles, theories, and methods of climate science is *not easy* and often involves simplifications (e.g. CO_2 often stands in for all greenhouse gases; the 2C/1.5C target is used as a threshold value), metaphors ("the greenhouse effect"), or the use of iconic, loaded images (say polar bears on a melting ice floe).

The advantages of such simplification are obvious. They allow for disseminating complex topics and feeding the findings of comprehensive observation and sophisticated experimenting to a wider public. But there are also disadvantages. Reducing a complex issue to simple, easy-to-grasp images, metaphors and numbers entails the risk that scientific findings get distorted or is misrepresented.

Perhaps the most prominent example for such a misrepresentation or misinterpretation respectively is the shift from "global warming" to climate change" as the most prominent label used in public and political discourses. As Conway (2008) points out, in the late 1980s, that is, in the run-up years to the Earth Summit in 1992, climatic change was usually referred to as "global warming". While the term became particularly prominent in early 2007, around the time when former US Vice President Al Gore released its popular documentary *An Inconvenient Truth*, it is increasingly replaced by the label "global climate change" (cf. Schuldt 2016: 7). The shift can be 'explained in large part by its greater scientific appropriateness and nuance (i.e., capturing the many diverse impacts of human activities, beyond rising temperatures, that influence climate patterns and biological systems at varying scales)' (ibid.).[10]

Theoretical Conception of the Knowledge: Decision-Making Relation

This leads to the important questions of which role can "knowledge" play in decision-making and how to conceptualize the role of scientific knowledge in climate politics? Moreover, answering these questions is crucial of getting a better understanding of the roles, strategies, and significance of think tanks in the process. Broadly speaking one can distinguish between three distinct conceptualizations of the relation between scientific knowledge and decision-making:

> First, scientific knowledge can be regarded as a transformative force ultimately leading to more "rational" decisions and "evidence" or "knowledge based" policymaking (Grundmann and Stehr 2012: 7). Second, more pragmatic approaches argue that that political decisions will be based on "incremental, negotiated solutions that work" (ibid.). That means that scientific knowledge has to immediately "useful", "comprehensible" or "applicable" (cf. Dilling and Lemos 2011) or it might be ignored, contested or rejected by political decision makers. Finally, scientific knowledge might be just another tool that can be used by policymakers to pursue certain political goals. Scientific knowledge will then only be accepted only if it allows the confirming, or legitimizing pre-existing normative political positions.

The first or transformative interpretation can be summed up in the straightforward, famous conception of "speaking truth to power" (Wildavsky [1979]2007; Jasanoff 1990). Such an understanding presupposes that "truth" can be fed into the political system and translated for and into political decisions. From this perspective, it is crucial that decision-makers can *understand* the content and the implications of scientific research. Normative convictions, ideological positions, and political preferences are, in the light of superior knowledge and irrefutable evidence, rendered irrelevant since '[a] science-based solution will be agreeable to warring parties, since it transcends the ideological [...] differences (Grundmann and Stehr 2012: 6). To take the metaphor further, proponents of such a model assume the political realm will not eventually learn to understand the language of science but will increasingly accept its primacy: Politicians, and the public, turn into receivers of the best scientific knowledge available on which they indirectly, or directly, base their decisions.

However, such conceptions can be criticized for several reasons. First, they tend to ignore some fundamental problems of the philosophy of science, most notable problems associated with concepts of "truth" or "certainty". As was demonstrated above even the best scientific knowledge of the causes and consequences of climate change isn't—and cannot be—100 per cent certain. Moreover, since climate science is, as was discussed, hardly accessible for people outside the scientific community, it is almost impossible to estimate the level of certainty. Policymakers and voters alike must decide whether to "trust" science an academic institution or organization or even a particular scientist, or not. Policymakers cannot wait until truth is spoken to them but are confronted by rivalling knowledge claim and it might not always be easy to tell scientific from non- or pseudoscientific claims.

Selling Doubt, Spreading Ignorance, and the Art of "Agnotology"

As Grundmann and Stehr (2005: 7–8) have pointed out, politicians will not happily surrender normative positions to the (alleged) superiority of scientific knowledge. From this perspective, policymakers will be less interested in truth but more interested in identifying "useful knowledge", that is, scientific knowledge that is backing political positions, agendas, or policies.

However, the straightforward and this more pragmatic model shares a common view on the *functionality* of scientific knowledge: From both perspectives, knowledge seems to be the opposite of "ignorance" understood as the 'absence or void where knowledge has not yet spread' (Proctor 2008: 2). The third analytical perspective mentioned above paints a much darker picture. In this view, "knowledge" cannot only be used but also abused to backing long-held normative views. Especially in democracies, where political decision-making needs to consider public sentiments, "deliberately engineered" ignorance (ibid.: 3) can serve as a "strategic ploy" to preserve the status quo and legitimate political non-decision. What if climate science "finds out" that something should be done about climate change but the proposed action conflicts with the political preferences of a political party? In this case "inaccurate" or deliberately "false" interpretation of knowledge claims translation can be in the interest of (some) policymakers: The manufacturing of doubt, the fostering of "impressions of implacable controversy where actual disputes are marginal," (Mirowski 2013: 227) can develop into a valuable service. The service could be labelled "applied agnotology" owing to the term "agnotology" introduced by Robert Proctor and Londa Schiebinger (2008).

The term agnotology was coined by linguist Iain Boal and refers to three different states of ignorance: (1) The native state understood as something inferior "knowledge growths out of" (Proctor 2008: 4); (2) ignorance as the result of selective choice—one opts to look for answers "here rather than there" (ibid.: 7)—and most important for our purposes; (3) ignorance as an active construct or, as mentioned above, as a strategic ploy (ibid.: 8–9).

"Agnotological" practices can be found in US American climate politics. Philip Mirowski (2013) provides empirical evidence that conservative policymakers and pressure groups align with suitable researchers and "experts" not to defeat climate science but to preserve the political status quo by prolonging public dispute. Mirowski speaks of a joint effort to 'muddy up the public mind and consequently foil and postpone most political action and (…) to preserve the status quo ante' (2013: 226). To prove his point, he recalls a memo of Frank Luntz to the Republican Party in the United States:

> The scientific debate remains open. Voters believe that there is no consensus about global warming. Should the public come to believe that the scientific issues are settled, their views about global warming will change accordingly. Therefore, you need to continue to make the lack of certainty a primary issue in the debates. (Luntz, cited in Mirowski 2013: 227)

Aaron McCright and Riley Dunlap have come to similar conclusions. In a series of studies, they showed how conservative think tanks and special interest groups managed to prevent pro-environmental decision-making in the US by challenging knowledge-based advice through the manufacturing of scientific dissent (McCright and Dunlap 2010: 105–106; also McCright and Dunlap 2011; Dunlap and McCright 2015; Jacques et al. 2008, Dunlap and Jacques 2013).

These examples show that the use of scientific knowledge isn't necessarily leading to the "rationalization" of policymaking nor to the reduction of ignorance.

Conclusion and Outlook: Towards Investigating the Role of Think Tanks

The three models described above are helpful in analysing the possible roles think tanks can play. Before developing a detailed theoretical and analytical model of think tanks (that takes into account the factors of political environment, diverging strategic goals and distinct "knowledge regimes"), it is important to ask what roles think tanks could fulfil.

In the straightforward conception of "speaking truth to power", think tanks would find their role reduced to functioning as gatekeepers, making sure that relevant knowledge becomes available to policymakers who then base their decisions on the best available knowledge and the most convincing evidence.

As we have seen above and as will be discussed in subsequent chapters, the straightforward conception of the relation between scientific knowledge and political decision-making is particularly widespread in climate science and climate politics. Accordingly, any behaviour or any strategy of think tanks that goes beyond "spreading the word" conflicts with the fundamental assumption of this perspective and therefore often a main subject of criticism.

The second and in particular the third perspectives, however, allow for a wider range of think tank activities. Think tanks can help in selecting or even manufacturing scientific knowledge claims that support specific political programmes. It is these possibilities that inform the rest of the book.

To summarize, this chapter has aimed at illustrating the particular role of scientific knowledge in establishing the claim that climate change is real and anthropogenic. Moreover, it shows that establishing certainty among climate scientists is a very different matter to convincing the public. Climate politics requires not a knowledge of climate change but rather a *belief in* it.

This is precisely where think tanks play a role and why their various services can be particularly valuable in the field of climate politics.

The following chapters outline an analytical framework that allows a more thorough analysis of what think tanks are doing, helping to address the question of how their strategies resonate with their respective institutional environment and to explain the persisting differences between think tanks in two selected countries, Germany and the United States.

Notes

1. From a legal perspective, *insanity* can be described as 'mental illness of such a severe nature that a person cannot distinguish fantasy from reality' (cf. http://dictionary.law.com/Default.aspx/Default.aspx?selected=979).
2. For a more detailed discussion, see Stehr and Machin (2018) *Climate & Society.*

3. Scientists had it all wrong, for instance, when they thought that the use of asbestos was harmless or that frontal lobotomies are an appropriate treatment for mental disorders.
4. Scepticism is arguably a cardinal virtue of the scientists, as, for instance, T.H. Huxley reminds us: 'For the improver of natural knowledge scepticism is the highest of duties, blind faith the one unpardonable sin' (Huxley, cited after Powell 2011: 4).
5. For a more detailed discussion of the 2C target, see Knopf et al. (2012).
6. The report of the United Nations "Brundtland Commission", named after its chair, the former Norwegian Prime. Minister Gro Brundtland was an important catalyst of global awareness and debate of environmental problems. Titled *Our Common Future*, the report is widely regarded as the first attempt to develop a comprehensive approach to sustainable development. It was further instrumental in setting up UN summits on environmental issues starting with the Earth Summit in Rio de Janeiro in 1992.
7. On its website, *The Telegraph* provides its readers a best of the leaked emails: http://www.telegraph.co.uk/news/earth/environment/globalwarming/6636563/University-of-East-Anglia-emails-the-most-contentious-quotes.html.
8. James Powell, for instance, refers to a highly controversial email in which the paper of a climate sceptic researcher is called "just garbage" and the sender promises to somehow keep it out of the upcoming IPCC Report. While the email seems to prove that the community of climate scientist is hopelessly biased and systematically suppresses dissenting views, paper rejections are actually part of standard scientific practice. Also, the mail doesn't prove that the authors intend to reject high-quality research for political reasons but rather that they doubt the quality of the research itself (cf. Powell 2011: 161–162).
9. Peer-reviewed processes and ethic commissions, for instance, shall make sure that scientific publications meet some quality and ethical standards.
10. Interestingly Schuldt finds that conservative, climate sceptic think tank in the USA predominantly make use of the term global warming, while liberal think tanks prefer "climate change" (Schuldt 2016: 21).

References

Arrhenius, Svante. 1896. On the Influence of Carbonic Acid in the Air Upon the Temperature of the Ground. *The London, Edinburgh and Dublin Philosophical Magazine and Journal of Science* 41: 237–276.

Chaitin, Daniel. 2017. Gov. Jerry Brown: Trump Fighting a Losing Battle with 'Reality' on Twitter. *Washington Examiner*, May 31. http://www.washington-

examiner.com/gov-jerry-brown-trump-fighting-a-losing-battle-with-reality-on-twitter/article/2624632. Accessed 3 June 2017.
Collins, Harry M. 1992. *Changing Order. Replication and Induction in Scientific Practice*. Chicago/London: Chicago University Press.
Conway, Eric. 2008. What's in a Name? Global Warming vs. Climate Change. https://www.nasa.gov/topics/earth/features/climate_by_any_other_name.html. Accessed 12 May 2008.
Cousins, Farron. 2017. House Science Committee Leader Says Climate Scientists Are Trying to Control People's Lives. desmogblog.com, April 7. https://www.desmogblog.com/2017/04/07/house-science-committee-lamar-smith-says-climate-scientists-trying-control-people-lives. Accessed 29 May 2017.
Dilling, Lisa, and Maria Carmen Lemos. 2011. Creating Usable Science: Opportunities and Constraints for Climate Knowledge Use and Their Implications for Science Policy. *Global Environmental Change* 21: 680–689.
Dunlap, Riley, and Peter Jacques. 2013. Climate Change Denial Books and Conservative Think Tanks. Exploring the Connection. *American Behavioral Scientist* 57 (6): 699–731.
Dunlap, Riley E., and Aaron M. McCright. 2015. Challenging Climate Change: The Denial Countermovement. In *Climate Change and Society: Sociological Perspectives*, ed. Riley E. Dunlap and Robert Brulle, 300–332. New York: Oxford University Press.
Edwards, Paul N. 2010. *A Vast Machine. Computer Models, Climate Data, and the Politics of Global Warming*. Cambridge, MA/London: The MIT Press.
Eurobarometer. 2009. Europeans' Attitudes Towards Climate Change Special Eurobarometer Report 313.
Fourier, Jopseph. [1824] 1936. General Remarks on the Temperature of the Terrestrial Globe and the Planetary Spheres. *American Journal of Science* 32(1): 1–20.
Grundmann, Reiner, and Nico Stehr. 2005. *Knowledge: Critical Concepts*. New York/London: Routledge.
———. 2012. *The Power of Scientific Knowledge. From Research to Public Policy*. Cambridge: Cambridge University Press.
Hansen, James. 2010. *Storms of My Grandchildren. The Truth About the Climate Catastrophe and Our Last Chance to Save Humanity*. London: Bloomsbury Publishing.
Hulme, Mike. 2009. The Origins of the 'Greenhouse Effect': John Tyndall's 1859 Interrogation of Nature. *Weather* 64 (5): 121–123.
IPCC. 2007. Climate Change 2007: Synthesis Report. In *Contribution of Working Groups I, II and III to the Fourth Assessment Report of the Intergovernmental Panel on Climate Change*, ed. Core Writing Team, R.K. Pachauri, and A. Reisinger, 104. Geneva: IPCC.

IPCC. 2013. Summary for Policymakers. In *Climate Change 2013: The Physical Science Basis. Contribution of Working Group I to the Fifth Assessment Report of the Intergovernmental Panel on Climate Change*, ed. T.F. Stocker, D. Qin, G.-K. Plattner, M. Tignor, S.K. Allen, J. Boschung, A. Nauels, Y. Xia, V. Bex, and P.M. Midgley. Cambridge, UK/New York: Cambridge University Press.

IPCC Fifth Assessment Synthesis Report – Climate Change 2014 Synthesis Report (1 November 2014) by Myles R. Allen, Vicente R. Barros, John Broome, et al. edited by Paulina Aldunce, Thomas Downing, Sylvie Joussaume, et al.

Jacques, Peter J., Riley E. Dunlap, and Mark Freeman. 2008. The Organisation of Denial: Conservative Think Tanks and Environmental Scepticism. *Environmental Politics* 17 (3): 349–385.

Jasanoff, Sheila. 1990. *The Fifth Branch. Scientific Advisors as Policymakers.* Cambridge, MA: Harvard University Press.

Jaspal, Rusi, and Brigitte Nerlich. 2012. When Climate Science Became Climate Politics: British Media Representation of Climate Change in 1988. *Public Understanding of Science* 23 (4): 122–141.

Keeling, Charles. 1960. The Concentration and Isotopic Abundances of Carbon Dioxide in the Atmosphere. *Tellus* 12: 200–203.

Knopf, Brigitte, Martin Kowarsch, Christian Flachsland, and Ottmar Edenhofer. 2012. The 2°C Target Reconsidered. In *Climate Change, Justice and Sustainability*, ed. O. Edenhofer et al. Dordrecht: Springer.

Kohn, Alexander. 1986. *False Prophets. Fraud and Error in Science and Medicine.* Oxford: Blackwell.

Lidskog, Rolf. 1996. In Science We Trust? On the Relation Between Scientific Knowledge, Risk Consciousness and Public Trust. *Acta Sociologica* 39 (1): 31–56. Sociology and the Environment (1996).

Machin, Amanda, and Alexander Ruser. forthcoming. What Counts in the Politics of Climate Change? Science, Scepticism and Emblematic Numbers. In *Science, Numbers Politics*, ed. Heidelberg Academy of Science. The paper will be published in Markus Prutsch (ed:) Working Numbers. palgrave

McCright, Aaron M., and Riley E. Dunlap. 2010. The Politicization of Climate Change and Polarization in the American Public's View of Global Warming 2001–2010. *The Sociological Quarterly* 52: 155–194.

———. 2011. Cool Dudes: The Denial of Climate Change Among Conservative White Males in the Unites States. *Global Environmental Change* 21: 1163–1172.

Miller, Norman. 2009. *Environmental Politics. Stakeholders, Interests, and Policymaking.* 2nd ed. New York/London: Routledge.

Miller, Clark A., and Paul N. Edwards. 2001. Introduction: The Globalization of Climate Science and Climate Politics. In *Changing the Atmosphere. Expert*

Knowledge and Environmental Governance, ed. Clark A. Miller and Paul N. Edwards. Cambridge, MA/London: The MIT Press.

Mirowski, Philip. 2013. *Never Let a Serious Crisis Go To Waste: How Neoliberalism Survived the Financial Meltdown*. London/New York: Verso.

Moser, Susanne C. 2010. Communicating Climate Change: History, Challenges, Process and Future Directions. *WIREs Climate Change* 1: 31–53.

Proctor, Robert. 2008. Agnotology. A Missing Term to Describe the Cultural Production of Ignorance. In *Agnotology. The Making and Unmaking of Ignorance*, ed. Robert Proctor and Londa Schiebinger, 1–36. Stanford: Stanford University Press.

Proctor, Robert N., and Londa Schiebinger. 2008. *Agnotology. The Making and Unmaking of Ignorance*. Stanford: Stanford University Press.

Powell, James Lawrence. 2011. *The Inquisition of Climate Science*. New York: Columbia University Press.

Randall, D.A., R.A. Wood, S. Bony, R. Colman, T. Fichefet, J. Fyfe, V. Kattsov, A. Pitman, J. Shukla, J. Srinivasan, R.J. Stouffer, A. Sumi, and K.E. Taylor. 2007. Climate Models and Their Evaluation. In *Climate Change 2007: The Physical Science Basis. Contribution of Working Group I to the Fourth Assessment Report of the Intergovernmental Panel on Climate Change*, ed. S. Solomon, D. Qin, M. Manning, Z. Chen, M. Marquis, K.B. Averyt, M. Tignor, and H.L. Miller. Cambridge, UK/New York: Cambridge University Press.

Revelle, R., and H. Suess. 1957. Carbon Dioxide Exchange Between Atmosphere and Ocean and the Question of an Increase of Atmospheric CO_2 During the Past Decades. *Tellus* 9: 18–27.

Ruser, Alexander. 2012. Greenhouse Gases. In *Encyclopedia of Global Studies*, ed. Helmut K. Anheier, Mark Juergensmeyer, and Faessel Victor, 774–775. Thousand Oaks: SAGE.

Ruser, Alexander. 2018. Experts and Science and Politics. In *The SAGE Handbook of Political Sociology*, ed. William Outhwaite and Stephen P. Turner, 767–780. Los Angeles: SAGE.

Schuldt, Jonathon P. 2016. 'Global Warming' Versus 'Climate Change' and the Influence of Labeling on Public Perceptions. In *Oxford Research Encyclopedia of Climate Science*, ed. Oxford University Press. Oxford: Oxford University Press.

Stehr, Nico, and Alexander Ruser. 2018. Climate Change, Governance and Knowledge. In *Institutional Capacity for Climate Change Response: A New Approach to Climate Politics*, ed. Theresa Scavenius and Steve Rayner, 15–30. London/New York: Routledge.

Stehr, Nico, and Amanda Machin. 2018. *Society & Climate*. Singapore: World Scientific Publishing.

Stokes, Bruce, Richard Wike, and Jill Carl. 2015. Global Concern About Climate Change, Broad Support for Limiting Emissions. PEW Research Centre, November 5.

Tauber, Peter. 1997. *Science and the Quest for Reality*. Basingstoke: Palgrave.

Treinish, Llyod A. 1984. A General Scientific Information System to Support the Study of Climate-Related Data. NASA Technical Memorandum 86152, September 1984, Goddard Space Flight Center, Greenbelt, Maryland.

Waldholz, Rachel. 2017. Beyond Believers and Deniers: For Americans, Climate Change is Complicated. *Alaska Public Media,* March 30. http://www.alaska-public.org/2017/03/30/beyond-believers-and-deniers-for-americans-climate-change-is-complicated/. Accessed 2 June 2017.

Watts, Jonathan, and Kate Connolly. 2017. World Leaders React After Trump Rejects Paris Climate Deal. *The Guardian*, June 2. https://www.theguardian.com/environment/2017/jun/01/trump-withdraw-paris-climate-deal-world-leaders-react. Accessed 25 June 2017.

Wildavsky, Aaron. 1979[2007]. *Speaking Truth to Power: The Art and Craft of Policy Analysis*. New Brunswick/London: Taylor & Francis.

Zimmerman, Michael. 1995. *Science, Nonscience, and Nonsense. Approaching Environmental Literacy*. Baltimore/London: The Johns Hopkins University Press.

CHAPTER 3

What Think Tanks Do: Towards a Conceptual Framework

Think tanks are important because the media believes they are important and the media believes in this importance because think tanks tell them they are.
(Hames and Feasey 1994: 233)

Despite having attracted to the interest of a growing number of political scientists and sociologists, thus "generating vast quantities of policy research" (Hird 2005: 1), think tanks remain rather mysterious players on the political stage (or behind the scenes). J.H. Snider observes that '[d]espite think tanks' billions of dollars of tax subsidies and considerable power, they have received minimal public scrutiny and are often poorly understood' (Snider 2009).

To be true, Sniders lament that think tanks have escaped public attention is not entirely true: Some think tanks, especially those in the USA, such as the famous Brookings Institution (founded in 1916 as Institute for Government Research) and RAND Corporation (established after World War II to advise the US military), for example, have gained considerable media attention. In his 1964 film *Dr. Strangelove*, renowned director Stanley Kubrick even set a cinematical monument to the RAND Corporation by modelling his protagonist after Herman Kahn (see Chap. 1).

Some think tanks seem to be influential because they are particularly visible. Conversely, think tanks are said to be influential precisely because they're invisible; these organizations can operate in the background,

A. Ruser, *Climate Politics and the Impact of Think Tanks*,
https://doi.org/10.1007/978-3-319-75750-6_3

providing discreet information to policymakers or infusing carefully tailored bits of information and interpretation to public debates, pulling the strings behind the scenes. Yet, regardless of their visibility, it is evident that the numbers of think tanks are increasing and that they are spreading globally (McGann 2016, 2017).

According to Donald Abelson, then, the global spread of think tanks is 'indicative of their growing importance in the policy-making process, a perception reinforced by directors of think tanks, who often credit their institutions with influencing major policy debates and government legislation' (2009: 3). So, although it remains unclear what precisely think tanks are or what exactly they do, they are said to be becoming more influential. Of course, this impression might be deliberately created or encouraged by representatives of think tanks.

It would be unfair, however, to blame researchers for failing to provide a comprehensive account of the outlook, behaviour, strategies, and impact of think tanks on the policymaking processes. To research these curious organizations one has to overcome enormous obstacles. For instance, since some think tanks look remarkably similar to more classical special interest groups and lobby organizations (McGann and Weaver 2000: 7), it could be in their own interest to maintain secrecy about some of their activities (e.g. sources of income, spending, or recruiting practices). Moreover, the term "think tank" itself somehow falsely implies a uniformity of structure and properties that belies the true diversity of organizations that are labelled by it. In the relevant literature, the term is applied to a diverse set of research institutions, public policy institutes, and consultancies. It has been described as more or less an 'umbrella term that means many different things to many different people' (Stone 1996: 9). For Thomas Medvetz, 'the basic problem is that the central concept itself is fuzzy, mutable and contentious' (2012: 23) and thus the definition of what a think tank is can change over time (Medvetz 2012: 26–28).

Despite a general 'lack of consensus (…) in defining think tanks' (McGann and Johnson 2005: 11), however, I use a working definition depicting them as 'independent, non-profit research facilities, engaged in applied research provided to political decision makers' (Ruser 2013: 331).

Perhaps the most elaborate definition was developed by Donald Abelson. He begins with a list of basic organizational characteristics that most think tanks have in common: 'they are generally nonprofit, nonpartisan organizations engaged in the study of public policy' (Abelson 2009: 9). Additionally, he considers the relatively broad spectrum of think tank behaviour:

> Think Tanks can embrace whatever ideological orientation they desire and provide their expertise to any political candidate or office-holder willing to take advantage of their advice. [...] Not all think tanks share the same commitment to scholarly research or devote comparable resources to performing this function, yet it remains, for many, their raison d'être. (Abelson 2009: 10)

However, defining a set of activities or listing organizational features is barely enough to fully understand what think tanks do, to single out their idiosyncrasies, and, most importantly, if and how they can influence policymaking in their respective political environment.[1] Nevertheless, Abelson's detailed definition offers a suitable starting point for developing a more comprehensive, conceptual framework, which combines variation at the organizational level (typologies), distinct points of intervention in public discourses (role and types of ideas), and the particularities of different institutional environments (knowledge regimes).

The conceptual framework that will be developed in the following paragraphs draws on Thomas Medvetz's insight that explaining think tank activities on the organizational level isn't enough; the 'forces and conditions that shape their practices' (2012: 35) have to be taken into account too. Still, however accurate, an analytical framework is never able to fully capture the ambiguity of public policy institutes: Boundaries between research institutes, advocacy organizations, or lobby groups are notoriously fuzzy. As will be discussed in more detail below, the term "think tank" itself may have a strategic value that may be used by organizations that want to distance themselves from lobby or special interest groups (see also Medvetz 2012: 34). Moreover, organizations do not need to commit themselves to a fixed set of techniques, strategies, or other organizational features but can behave more or less "think-tankishly" and switch between different roles if this serves their particular goals.

A Snapshot of a Frustrating Venture: The State of Think Tank Research

Given the ambiguity of the object of study, it is not surprising that scholarly work on think tanks is still in an infant state and 'few definitive answers [have] been offered' (Abelson 2009: 3). What is surprising, however, is the widespread view (McGann and Weaver 2000: 5ff; Stone 2004: 1f) that think tanks are influential no matter how impenetrable they are.

Twenty-five years ago David Ricci remarked that '[p]ower in Washington cannot be measured precisely, yet think tanks surely have a good deal of it' (1993: 2). It seems little has changed. Murray Weidenbaum, for instance, more recently acknowledges that think tanks might become more important players on the political arena, conceding at the same time that '[a] ctually trying to measure their impact on specific public policy changes, however, has frustrated scholars for years' (2010: 135). Weidenbaum, a former chairman of the council of economic advisors to Ronald Reagan and intimately acquainted with political Washington, describes the undertaking to provide a precise measurement of think tank influence as akin to the difficult task of identifying pornography: Instead of defining it, one has to limit oneself to acknowledging: 'I know it when I see it' (Weidenbaum 2010: 135).

Richard Cockett (1994) and Keith Dixon (1998) certainly seemed to 'know it when they saw it', conclusively demonstrating the influence of right-wing think tanks in the making of an "economic counter revolution" and the rise of Thatcherism. For Philip Mirowski (2013), too, think tanks belonged to the inner circles of a neoliberal "thought collective" (2014: 44) that successfully defended their preferred economic and political programme in spite of the massive financial crises since 2007.

With regard to the contested issue of anthropogenic climate change, the potential influence of think tanks on the decision-making process has drawn even more attention: In a series of empirical studies, Aaron McCright and Riley Dunlap (2003, 2010, 2011, 2015) were able to show how conservative think tanks contributed to fostering "climate denial" in the United States. Naomi Oreskes and Erik Conway's seminal monograph *Merchants of Doubt* (2011) traces the impact of think tanks on a variety of public policy issues such as global warming, acid rain, and the carcinogenic effects of smoking.

Notwithstanding these accounts, analysing and "measuring" think tank influence is a delicate undertaking. It is possible to measure the "output" of think tanks, for instance, by counting publications such as policy briefs, books, or op-eds. Attempts to quantify think tank influence include James McGann's Global Go To Think Tank Index Reports (2010, 2016; McGann and Sabatini 2011) which focus predominantly on basic quantitative data (e.g. the number of think tanks in a given country, or the growth in numbers over time) or expert estimation (subjective ranking of a respective think tank's impact in a particular country and/or issue area). But tracing the influence of think tanks directly to specific paragraphs or concepts in a specific piece of legislation is very difficult if not impossible.

Yet the very existence of a "global" think tank survey indicates a major transition in research on think tanks. For a long time these organizations had been described as a specific phenomenon of the political systems of Anglo-Saxon countries, in particular, the United States. This view was prominently formulated in David Ricci's "habitat hypothesis":

> What needs explaining [...] is not how to build and run a think tank but why so many of them became prominent in Washington during the 1970s and 1980s, more than was the case in, say, Rome, Ankara, Riyadh, Bonn or Djakarta. Here, what mattered was especially democratic context rather than vocational substance. We need to know, for example, why Washington's political people became responsive to think-tank products even though those were, after all, standard research items other countries or eras might have ignored. (Ricci 1993: 3)

From this perspective, the assessment made by McGann two decades later that 'think tanks now operate in a variety of political systems, engage in a range of policy-related activities and comprise a diverse set of institutions that have varied organizational forms' (McGann 2010) demands further explanation. Does it suggest a catching-up development with the USA on the global scale? At first glance, this seems to be a plausible explanation since, discounting national differences, 'most think tanks share a common goal of producing high quality research and analysis that is combined with some form of public engagement' (Braml 2006: 223). Maybe modern societies become more responsive to applicable research and analysis, rendering the global spread of think tanks a demand-driven adjustment in the market for political consulting.

To answer this question, we have to investigate what think tanks are actually doing. Can we arrive at a consistent, uniform description of think tank activities? Or should the social and political particularities of different institutional environments think tanks operate in be taken into account?

For instance, some think tanks are doing research and providing scientific expertise to policymakers and/or the wider public. As we will see below, it is tempting to focus on such features to develop think tank typologies. Think tanks with research capacities of their own could then be labelled "academic" think tanks, while think tanks without such capacities would be labelled differently. We could than compare the respective "mix" of think tanks in various countries. However, this might not be enough, for it would neglect how research conducted by think tanks would "fit in" national research systems. We therefore should take into account the respective relation of think tanks to academic organizations such as universities, public research laboratories, corporate research units, or privately funded research institutions.

Another important dimension is the public engagement of think tanks. In order to understand country-specific think tank behaviour, it is necessary to analyse the distribution of specific techniques to approach the public. Comparative studies look for specific patterns or "tradition" of think tank activities (cf. Stone and Denham 2004) or investigate the impact of specific institutional settings on the strategies pursued by think tanks (Ruser 2013; Sheingate 2016). Comparative studies fuel doubt as to whether the global spread of think tanks can be explained by a "catch-up development". It seems increasingly more plausible to assume parallel developments indicate that think tanks may fill different roles in different societal and political environments.

Furthermore, global trends towards "knowledge societies" cannot explain directly the increasing number of think tanks worldwide. It is unlikely that institutional differences between societies will vanish in the process of transformation (Böhme and Stehr 1986: 13–14) as is the convergence of the ways knowledge is fed in processes of public deliberation and decision-making.

Think Tanks and Social Capital: Analysing Think Tanks Within Their Networks

Defining what a think tank actually doing is difficult because of the organizational diversity, different institutional environments, and diverging strategies for exercising influence.

Things are further complicated by the fact that think tanks are not the only organizations that seek to provide expertise and influencing political debates. Tom Medvetz therefore argues in favour of a 'clear break from both ordinary and scholastic common sense definitions of this term, which generally try to locale its meaning in a particular essence, substance, or population of organizations' (Medvetz 2012: 34–45).

Instead he proposes a relational definition (ibid.) that includes the 'institutional positions of the organizations that acquire this label and forces and conditions that shape their practices' (Medvetz 2012: 35). Medvetz employs French sociologist Pierre Bourdieu's concepts of "social space" and "fields of power" to *locate* organizations labelled think tanks and describing them by referring to Bourdieu's "forms of capital" (1986). Medvetz's approach is valuable for two reasons. First, it turns the task of theoretically defining think tanks into an empirical question. Second, instead of aiming for a rigid typology of think tanks, Medvetz's concept more flexibly includes "hybrid" organizations (i.e. organization that act more or less think-tankishly) (ibid.: 36.)

However, the model also harbours some difficulties, in particular, for comparative studies. First, the use of Bourdieu's forms of capital can cause serious empirical problems. Bourdieu distinguishes between three forms of capital: economic, cultural, and social. Most straightforwardly, economic capital refers to all kinds of economic resources (e.g. income, assets, or property) and can be easily applied to individuals (to measure the "wealth" of an individual) and organizations (e.g. the budget). Economic capital of think tanks is therefore easy to measure. But measuring the amount of cultural and social capital to such organizations is considerably more difficult.

Cultural capital can be *embodied*, *objectified*, and *institutionalized* (Bourdieu 1986). Since embodied cultural capital refers to individual knowledge, but also a certain taste, communication skills, and linguistic abilities, it cannot be applied directly to organizations. Certainly, organizations can "stand for" a set of values, educational principles, or other components of cultural capital, it's the members of the organization who have to embody the cultural capital. Organizations can have the other forms of cultural capital: objectified cultural capital which refers to material objects (e.g. works of art, a library, musical instruments) and institutionalized cultural capital, which refers to more immaterial objects (formal qualification and certificates such as college diplomas but also certified memberships, as in the case of an "ivy league" university). However, estimating organizational cultural capital empirically is notoriously difficult. One reason for this is the exchange rate between individual and organizational cultural capital. Organizations can lend their cultural capital to individuals and vice versa. Becoming a professor at a prestigious university, working at the White House, or being employed by a leading company can increase one's cultural capital. On the other hand, a small university or a start-up lacking reputation might hire "big names" in order to gain cultural capital (and hence transforming economic capital into cultural capital). But hiring big names, maintaining close ties to political parties, or being a member in a professional network might increase an organization's cultural capital in one context (say one country) but might decrease cultural capital in another. Thus, the value of cultural capital can only be estimated relative to its specific setting. This makes it particularly difficult to compare the cultural capital of organizations from different settings.

Since the conceptual framework that is developed in this chapter aims at an international comparison of think tanks, it therefore deviates from the Medvetz approach. Instead of focusing on cultural capital, it will focus upon their social capital, the term Bourdieu used to emphasize the possession of/position within a social network:

> Social capital is the aggregate of the actual or potential resources which are linked to possession of a durable network of more or less institutionalized relationships of mutual acquaintance and recognition – or in other words, to membership in a group – which provides each of its members with the backing of the collectivity-owned capital a 'credential' which entitles them to credit, in the various senses of the word. (Bourdieu 1986: 51)

To analyse the volume and, more importantly, the value of social capital, it is appropriate to investigate the size and the structure of social networks. Social capital provides its owner with *access* to information, support, and financial resources and allows them to *reach* other members within the network. Moreover, the network perspective allows for the analysis of whether (political) actors 'reside in homogenous social settings (…) and hence the information they obtain is a direct reflection of their own political biases' (Zuckerman 2005: 23): Social capital, displayed in the structure of social networks, therefore determines what information and what interpretation of information are available to members of a given network.

However, the size and the shape the networks itself is affected and constrained by a complex interplay of other forms capital. Access might be limited by economic constraints (e.g. if one has to pay for a campaign add), but network member might be out of reach for other reasons. "Established" organizations, for instance, "venerable universities", might prevent "upstarts", for example, newly established research institutions from getting access to expert panels and other important committees.

This channelling of "resources, communications, influence, and legitimacy" within a social network in turn helps create or foster 'shared identities and collective interests, and thus promote the acceptance of a common field frame' (Knoke and Yang 2006 cited after Brulle 2014: 689).

The central assumption therefore is that the roles think tanks play and the strategies they pursue can be identified by analysing their respective networks. Whom do they cooperate with and which actors or fields are out of reach? Moreover, it has to be analysed how these networks are structured and constrained by external factors.

However, to keep the analytical model simple enough to actually apply it to empirical cases, it has to rely in older conceptualizations of think tank typologies.

Think Tank Typologies as Heuristics

Although Thomas Medvetz rightly criticizes classical think tank typologies, they can nevertheless be useful to derive testable hypotheses of how think tanks can act within larger networks. Typologies can therefore be used as *heuristics* that inform rather than limit the analytical model.

Kent Weaver was among the first who tried to classify think tanks. His typology published in 1989 takes into account different patterns of recruitment, funding, output, and audience of think tanks in the United States. He then differentiates between "universities without students" (UWS) and advocacy think tanks. UWS 'tend to be characterized by heavy reliance on academics as researchers, by funding primarily from private sector [...], and by book-length studies as the primary research product' (Weaver 1989: 564). The purpose of this type of think tank is to provide scientific advice to their clients and contribute to shape the "climate of elite opinion" (Weaver 1989: 564). UWS as well as "contract research organizations" (ibid.: 566)—which compile scientific reports for government agencies/contractors—maintain the high standards of academic inquiry and can be labelled "academic think tanks".

According to Weaver, think tanks differ with regard to their preferred activities, strategic behaviour, and staff employed. To further examine how academic and advocacy think tanks can exercise influence, one can refer Thomas Osborne's (2004) model of different types of political influence. Osborne distinguishes between two ideal types: First, an advisory or *leverage model* in which personal or professional reputation is directly translated into public credibility and subsequently into political authority (Osborne 2004: 433–434), and second, a "brokerage model" in which influence is exercised if 'vehicular ideas [are] brokered between parties, designed to enhance particular kinds of outcome' (ibid.: 434).

Hence, it can be assumed that academic think tanks exercise influence according to the "leverage model". The (individual) prestige of its researchers, adherence to established rules for scientific research, and the authority of science itself allow academic think tanks to play the role of impartial experts, claiming to speaking "truth to power".

In contrast, advocacy think tanks 'combine a strong policy, partisan or ideological bent with aggressive salesmanship and an effort to influence current policy debates' (Weaver 1989: 567). Weaver's depiction suggests that advocacy think tanks, in contrast to academic think tanks, should employ a brokerage model.

Moreover, according to Weaver (1989: 568–569), think tanks can be a "source of policy ideas" function as evaluators of policy proposals and programmes, provide skilled personnel, and be a source of "punditry" for the media; it is necessary to distinguish between the more passive provision of knowledge and a more active engagement in the translation process.

Especially since scientific knowledge has to be translated in order to be "usable" and "useful" (see Chap. 2), scientific experts have to decide how to engage in this translation process. According to Roger Pielke Jr., the scientific expert has to choose between four idealized roles: the pure scientists, the science arbiter, the issue advocate, and, finally, the honest broker. The pure scientist would recuse herself from the process and 'focus on research with absolutely no consideration for its use or utility, and thus in its purest form has no direct connection with decision-makers' (Pielke 2007: 15). Like the pure scientist, the science arbiter distances herself from political considerations. Unlike the pure scientist, however, science arbiters frequently interfere directly with political decision-makers and 'focus on issues that can be resolved by science'.

In contrast, issue advocates become directly involved in the translation process by aligning themselves with specific social/political groups. Issue advocate adopt an active role in promoting political agendas, which are in accordance with the (perceived) implications of their research. Pielke's preferred role is certainly the "honest broker". The honest broker 'engages in decision-making by clarifying and, at times, seeking to expand the scope of choices available to decision-makers' (Pielke 2007: 17) and is eager 'to integrate scientific knowledge with stakeholder concerns in the form of alternative possible courses of action' (ibid.).

Combining Osborne's model and Pielke's typology of experts thus allows for refining Weaver's initial typology. The difference between UWS and advocacy think tanks can be described as a double continuum (Table 3.1):

Table 3.1 Ideal types of |academic and advocacy think tanks

Academic think tank	*Advocacy think tank*
Leverage model	Brokerage model
Science arbiter	Issue advocate

Source: Own research

> On the one end the pure academic think tank would rely exclusively on the personal reputation of its researchers (leverage model) who in turn adopt the role of science arbiters to preserve their professional integrity. On the other end of the continuum, the ideal-type advocacy think tank employs issue advocates that actively seek to broker between different groups in order to increase the impact of their expertise.[2]

However, it must be stressed that the scientific character (or at least a semblance of it) takes the centre stage for *all* types of think tanks. While all claim to speak with scientific authority, the attitude towards the standards of academic research varies considerably between academic and advocacy think tanks (cf. Weaver 1989; Stone 2004; Kern and Ruser 2010). Taken to the extreme, "advocacy" could be defined as readings to balance the adherence to the standards of good scientific practice against the principles of pre-existing normative ideas.

A Framework for Analysing Think Tank Behaviour

To assess the impact of think tanks, it is not sufficient to focus on organizational differences though. It is essential to take into account the context in which they operate. This in turn requires a more sophisticated model of political discourses and the role scientific knowledge, ideas, and expertise can play. This means that a comprehensive framework should focus on the respective institutional setting which channels access to decision-makers or the media and, more generally speaking, "sets the tone" for the provision of knowledge and consulting.

John Campbell's conceptualization of different types of ideas (1998) and their possible effects on policymaking can meet the first requirement. John Campbell's and Ove Pedersen's model of "knowledge regimes" (2011, 2014) provides a suitable starting point for the second, a systematic investigation of respective institutional context think tanks have to operate in.

Diane Stone describes think tanks as 'switchboards through which connections are made' (Stone 1996: 95). This depiction fits in the understanding of think tanks as translators but simultaneously begs the question of which connections are established. The notion of "bridging the gap" between knowledge and politics can refer to the process of political consulting in the literal sense. Knowledge is provided for policymakers to shape political action. This connection can either be straightforward, that is policies are directly shaped along experts' advice, or represent a more general understanding of how things should be done shared by policymakers and the members of the scientific community. Alternatively, the bridging of the knowledge gap may involve a form of public engagement. The provision of expertise can aim at helping people to "make sense" of complex situations and to sort out what not so much what is "thinkable" or "appropriate" or "acceptable".

John Campbell's "typology of ideas" (1998) allows distinguishing between these different forms of linking knowledge and politics. Table 3.2 presents four distinct effects of political ideas. Such ideas, understood as the content of a translation of knowledge for politics, can influence decision-making as concepts in the foreground or as underlying assumptions in the background of policy debates. Moreover, they can be located at the cognitive or normative level. At the cognitive level, they can function either as *programmes* (foreground), that is, they serve as policy prescriptions for the political elite necessary to formulate actual agendas, or as *paradigms* (background). In the latter case, they define the boundaries of "thinkable solutions" for political decision-makers. Likewise, at the normative level, they can provide frames suitable for legitimizing policy solutions or public sentiments constraining the range of legitimate solutions (Table 3.2).

Following John Campbell's model allows for deriving some hypothesis of think tank strategies and linking them to the respective type of think tanks and their target audience. As was discussed in Chap. 2, translating complex scientific knowledge has to be concerned with public opinion and public sentiment.

Influencing the "public sentiment" is an indirect yet powerful way of affecting political debate and political agendas. Producing content suitable for being used by the mass media, for instance, might fall short of catching the full complexity of the scientific research. It might nevertheless be

Table 3.2 Types of ideas and their effects on policymaking

	Concepts and theories in the foreground of the policy debate	*Underlying assumptions in the background of the policy debate*
Cognitive level	*Programmes* Ideas as elite policy prescription that help policymakers to chart a clear and specific course of policy action	*Paradigms* Ideas as elite assumptions that constrain the cognitive range of useful solutions available to policymakers
Normative level	*Frames* Ideas as symbols and concepts that help policymakers to legitimize policy solutions to the public	*Public sentiments* Ideas as public assumptions that constrain the normative range of legitimate solutions available to policymakers

Source: Campbell (1998: 385)

extremely effective since it can affect what is considered legitimate or morally acceptable. Think tanks pursuing this strategy will be engaged in writing op-eds and newspaper articles or sending their staff to appear on television, radio, or other channels. Also, think tank staff will not be limited to researchers (who might get "medial training") but can include PR specialists and journalists.

Likewise, it can be assumed that academic think tanks ("universities without students") will be less likely to engage with influencing public sentiment. The emphasis on scientific research and the necessity to meet the standards of good scientific practice bind considerable resources, and the simplification involved could undermine the integrity and credibility of its staff.

With regard to the think tank typology outlined above, it can be assumed that academic think tanks will most likely try to provide assistance in the formulation of programmes and frames. This hypothesis can be further refined when it comes to contracted (demand driven) research. Existing paradigms and public sentiments influence what clients consider either useful or legitimate and constrain the scope and the aim of an offered contract accordingly. A contracted academic think tank should serve its clients' needs best by providing evidence in support of a favoured political paradigms. The comprehensiveness of their research makes it less suitable for charting specific political action but may affect the range of solutions considered useful by politicians. In contrast, advocacy think tanks will primarily attempt to exert influence on the normative level. By providing expertise to legitimize policy solutions, these organizations can create or alter frames available to policymakers and influence public sentiments in a way that the solutions offered by advocacy think tanks are considered legitimate.

So far the emphasis was on the think tanks as organizations. So far the impact of the environment they operate in has been neglected. In the next step, the "interaction between ideas and institutions" (Campbell and Pedersen 2011: 167), that is, the influence of the "environmental conditions" that impact the prospects of different think tank strategies, has to be included.

The concept of "knowledge regimes" developed by John Campbell and Ove Pedersen provides a suitable starting point. The initial aim of the knowledge regime concept is to 'better understand how policy-relevant knowledge is created' (Campbell and Pedersen 2011: 168) by combining insights form research on policymaking and comparative studies on production regimes.

Campbell and Pedersen draw on the "varieties of capitalism" approach by Hall and Soskice (2001) to distinguish two types of production regimes: "Liberal market economies", structured by markets and corporate hierarchies on the one hand, and "coordinated market economies", on the other. In the latter case, economic activities are not dominated by the rules of free markets but are embedded in elaborate systems of corporatist bargaining and regulation (including state intervention) (Campbell and Pedersen 2011: 170). With regard to policymaking regimes, they follow Katzenstein's (1978) distinction between "centralized, closed states" (CCS) and "decentralized, open states" (DOS). Policymaking in CCS is located in few arenas, largely insulated from external influences. In contrast policymaking in DOS is more exposed to external influence, as authority is likely to be shared or delegated. By combining these two strands of theoretical thought, Campbell and Pedersen get four ideal-type knowledge regimes [see table below]. These knowledge regimes differ with regard to the overall impact of ideas (competitive vs. consensus-oriented systems) and in terms of which types of research units and think tanks being influential (cf. Campbell and Pedersen 2011: 171–172) (Table 3.3).

Table 3.3 Typology of knowledge regimes

	Liberal market economy	*Coordinated market economy*
Decentralized, open state	*Market-oriented knowledge regime* Large, privately funded research unit sector in civil society Scholarly and advocacy research units dominate Highly adversarial, partisan, and competitive knowledge production process	*Consensus-oriented knowledge regime* Moderate, publicly funded research unit sector in civil society Scholarly, party, and state research units even balanced Consensus-oriented, relatively nonpartisan knowledge production process
Centralized closed state	*Politically tempered knowledge regime* Small, publicly and privately funded research unit sector in civil society Scholarly, advocacy, and state research units evenly balanced Moderately adversarial, partisan, and competitive knowledge production process	*Statist technocratic knowledge regime* Large, publicly funded research unit sector in civil society Scholarly and state research units dominate Technocratic, nonpartisan knowledge production process

Source: Campbell and Pedersen (2014)

Habitat and Adaption: Different Institutional Contexts for Think Tanks in the USA and Germany

The two cases dealt with in this study can now be assigned to a specific knowledge regime type. According to Campbell and Pedersen, 'the US knowledge regime is highly competitive; (…) Germany's is coordinated' (2015: 2).

The competitive political system of the United States has 'facilitated the access of think tanks to various stages of the policy-making process' (Abelson 2009: 62). Indeed, according to Donald Abelson, 'few other countries provide an environment more conducive to the development of think tanks' (ibid.: 63). The division of power between the Senate and the House of Representatives filled representatives who are hardly bound by party unity, an increasingly polarized two-party system (Kuo and McCarty 2015: 49), and the overall importance of private donation provide think tanks not only opportunities 'to influence policy-making but has some ways increased the demand for them' (Abelson 2009: 64). For Andrew Rich, this increased demand indicates a shift in policymaking in the United States that favours "sleekly styled marketing machines" (2004: 208) which are gradually replacing the "neutral expert" (ibid.: 210).

The polarization of the political system (see also Chap. 4) was accompanied by an increasing significance of private donations and wealthy donors, who were particularly active in promoting conservative and free-market ideas, thus transforming the marketplace into a battlefield of ideas (Mayer 2016: 27ff).

Summing up the main features of and the most important trends within the US political system, Campbell and Pedersen describe the USA as follows:

> The USA is typically characterized as a liberal market economy with an institutionally porous and pluralist state rooted in a two-party, winner-take-all electoral system that facilitates divisive politics and lobbying by private interests (…).It is also a political system prone to enormous infusions of private money from a wide array of decentralized, independent interests, such as corporations, unions, wealthy foundations and wealthy individuals. (Campbell and Pedersen 2015: 7)

Accordingly, the United States can be described as a *market-oriented knowledge regime*, that is, as a "marketplaces for ideas" (where it literally pays off to pay for think tank advice). Competition dominates the process of knowledge production and distribution, enabling consultancies to engage in agenda-setting and the provision of support for political positions. Selling one's ideas is important and creates advantages for advocacy organizations engaged (and staffed with the necessary "issue advocates") in "aggressive salesmanship" (Weaver 1989: 567).

Although Germany 'began moving toward a somewhat more pluralist and competitive political economy' (Campbell and Pedersen 2015: 14), since the 1980s, it is still a decentralized, consensus-oriented system in which the 'state plays an important coordinating role' (ibid.).

Compared to marketplaces of ideas, such a *consensus-oriented knowledge regime* is a rather hostile environment for independent think tanks, in particular, advocacy think tanks. Here, the provision of expertise is guided by the principle of impartiality and is moderated by an overall orientation towards political consensus. Instead of asking whether 'Germany's marketplace of ideas [will] ever resemble America's' (*The Economist*, cited in

Braml 2006: 223), one has to analyse the impact of a specific "knowledge regime" on the production and distribution of scientific knowledge on the process of political decision-making and public discourses alike. Accordingly, the spread of think tanks has to be studied with regard to the specific rules and institutions that shape their organizational structures and strategies for exerting influence.

Moreover, the German political system is characterized by a strong consensus orientation (Katzenstein 1978). According to Kenneth Dyson, the Germany "policy style" is a combination of an "intellectual style" which values impartial, "objective," and scientific information and a "negotiation style" which emphasizes the importance of consensus to arrive at stable, long-term policy solutions (Dyson 1982 cited after Beuermann and Jäger 1996: 1987).

The importance of consensus is further increased by the need to form political coalitions. As a Federal Republic, Germany has one federal and 16 regional parliaments, all of which have coalition governments in place.

The basic law further introduces a principle of subsidiarity to the law-making process. According to Article 70(1), 'The *Länder* shall have the right to legislate insofar as this Basic Law does not confer legislative power on the Federation.' While the regional states are passing the majority of laws, Article 72 sets out the procedures for "concurrent legislation" and Article 73 lays down the '[m]atters under exclusive legislative power of the Federation'. In fields subject to concurrent legislation regional states so long and to the extent that the federation hasn't made use of its own legislative powers. The issue areas subject to concurrent legislation are listed in Article 72(3) of the basic law and include the "protection of nature and landscape management".[3]

The division power between the federal and the regional levels as well as the need to form coalitions favours political compromise and renders the "aggressive salesmanship" of advocacy think tanks presumably less relevant. Since coalitions may change, politicians have strong incentives to find "common ground" with other parties. While this tendency is sometimes criticized for blurring the boundaries between political parties and depriving voters of true political alternatives (Wiedemann et al. 2013: 81), the pragmatic approach has allowed political parties to adjust to a changing party landscape and voters' preferences. Even a coalition between the Christian Conservative Party and the Green Party which in the 1980s had been decried as "eco-fascist" (notably by high ranking politicians of the Social Democrats, Weissbach 1984: 32) became possible. Explaining how and why a coalition government between the Green Party and the Christian Conservatives could be forged in the regional state of Baden-Württemberg in 2016, political journalist Volker Wagener points exactly to the political/ideological flexibility of the parties involved:

> The Greens have become less fundamentalist and utopian, and much more pragmatic. But the CDU, too, is now more tolerant, it has adopted environment-friendly policies and has come to recognize that this approach need not hamper economic development. So both parties can team up to promote green business to boost the economy. (Wagener 2016)

Conclusion and Outlook: Analysing Think Tanks in Their Natural Habitat

The analytical framework outlined in the previous paragraphs serves one single purpose: It enables the analysis of American and German think tanks in their respective "natural habitat". The concept incorporates the objection made by Thomas Medvetz to study think tanks on the organizational level only. Instead, the focus of analysis will be on the social capital, that is, the position of think tanks within networks. In order to locate think tanks within a network, it is important to measure who think tanks are trying to reach and what kind of service they are seeking to provide.

But in order to be able to pinpoint the position of a think tank in a network, think tank typologies are needed. Despite various shortcomings, (sophisticated) think tank typologies can nevertheless serve as heuristics which contribute to this aim. John Campbell's model of the influence of ideas allows us to distinguish between different target audiences and to locate think tank influence on both the cognitive and the normative levels. Finally, the use of the regime typology developed by John Campbell and Ove Pedersen makes it possible to take external constraints into account. The networks within which think tanks operate in are pre-shaped by certain features of the respective political system, traditional means of knowledge production and private and public spending.

In the next two chapters this analytical model will be applied to analyse the role, the strategies, and the impact of think tanks on national climate politics in the United States (Chap. 4) and Germany (Chap. 5).

Notes

1. Relying on organizational aspects like non-profit or nonpartisan orientation is further complicated by diverging legal definitions of these very terms. Being labelled a "non-profit organization", for instance, has considerable influence on the tax burden of an organization. This explains why most American think tanks are so called 501(c) organizations, that is, they are exempt from federal income tax. German non-profit organizations also get favourable tax allowances but differ otherwise form their American "counterparts" (see Winheller 2010).
2. A good example is the description Edwin Feulner, president of the Heritage Foundation, a conservative, Washington-based think tank, gives of his organization. According to Feulner, Heritage is 'trying to influence the Washington public policy community (…) most specifically the Hill, secondly the executive branch, thirdly the national news media' (Abelson 2004: 220).
3. Article 74 of the basic law lays down additional matters subject to concurrent legislation including climate relevant issues such as "waste disposal, air pollution, and noise abatement", "management of water resources", "preservation of the coasts", and "meteorological services".

References

Abelson, Donald E. 2004. The Business of Ideas: The Think Tank Industry in the USA. In *Think Tank Traditions. Policy Research and the Politics of Ideas*, ed. Diane Stone and Andrew Denham. Manchester/New York: Manchester University Press.

———. 2009. *Do Think Tanks Matter? Assessing the Impact of Public Policy Institutes.* Montreal: McGill-Queens University Press.

Beuermann, Christiane, and Jill Jäger. 1996. Climate Change Politics in Germany: How Long Will any Double Dividend Last? In *Politics of Climate Change. A European Perspective*, ed. Tim O'Riordan and Jill Jäger, 186–227. London/New York: Routledge.

Böhme, G., and N. Stehr. 1986. *The Knowledge Society: The Growing Impact of Scientific Knowledge on Social Relations.* Dordrecht: D. Reidel.

Bourdieu, P. 1986. The Forms of Capital. In *Handbook of Theory and Research for the Sociology of Education*, ed. J. Richardson, 241–258. New York: Greenwood.

Braml, J. 2006. U.S. and German Think Tanks in Comparative Perspective. *German Policy Studies* 3 (2): 222–267.

Brulle, Robert J. 2014. Institutionalizing Delay: Foundation Funding and the Creation of U.S. Climate Change Counter-Movement Organization. *Climate Change* 122 (4): 681–694.

Campbell, John L. 1998. Institutional Analysis and the Role of Ideas in Political Economy. *Theory and Society* 27 (3): 377–409.

Campbell, J.L., and O.K. Pedersen. 2011. Knowledge Regimes and Comparative Political Economy. In *Ideas and Politics in Social Science Research*, ed. D. Béland and R. Cox, 167–190. Oxford: Oxford University Press.

Campbell, John L., and Ove K. Pedersen. 2014. *The National Origins of Policy Ideas. Knowledge Regimes in the United States, France, Germany, and Denmark.* Princeton/Oxford: Princeton University Press.

———. 2015. Policy Ideas, Knowledge Regimes and Comparative Political Economy. *Socio-Economic Review* 2015: 1–23.

Cockett, R. 1994. *Thinking the Unthinkable: Think Tanks and the Economic Counter-Revolution, 1931–1983.* London: HarperCollins.

Dixon, K. 1998. *Les évangélistes du marché: Les intellectuelles britannique et le néo-libéralisme.* Paris: Raison d'agir.

Hall, P.A., and D. Soskice. 2001. *Varieties of Capitalism: The Institutional Foundation of Comparative Advantage.* Oxford: Oxford University Press.

Hames, Feasey 1994: 233. Anglo-American Think Tanks Under Reagan and Thatcher. In *A Conservative Revolution? The Thatcher-Reagan Decade in Perspective*, ed. Andrew Adonis and Tim Hames. Manchester: Manchester University Press.

Hird, J.A. 2005. *Power, Knowledge and Politics.* Washington, DC: Georgetown University Press.

Katzenstein, P.J. 1978. *Between Power and Plenty: Foreign Economic Policies of Advanced Industrial States.* Madison: University of Wisconsin Press.

Kern, T., and A. Ruser. 2010. The Role of Think Tanks in the South Korean Discourse on East Asia. In *Korea 2010. Politics, Economy and Society*, ed. R. Frank, J.E. Hoare, P. Köllner, and S. Pares, 113–134. Leiden: Brill.

Kuo, Didi, and Nolan McCarthy. 2015. Democracy in America, 2015. *Global Policy* 6 (S1): 49–55.

Mayer, Jane. 2016. *Dark Money. The Hidden History of the Billionaires Behind the Rise of the Radical Right.* New York/London/Toronto/Sydney/Auckland: Doubleday.

McCright, A.M., and R.E. Dunlap. 2003. Defeating Kyoto: The Conservative Movement's Impact on U.S. Climate Change Policy. *Social Problems* 5 (3): 348–373.

———. 2010. Anti-reflexivity: The American Conservative Movement's Success in Undermining Climate Science and Policy. *Theory, Culture, and Society* 27 (2–3): 100–133.

McCright, A.M., and Riley E. Dunlap. 2011. Cool Dudes: The Denial of Climate Change Among Conservative White Males in the United States. *Global Environmental Change* 21: 1163–1172.

McCright, A.M., and R.E. Dunlap. 2015. Challenging Climate Change: The Denial Countermovement. In *Climate Change and Society: Sociological Perspectives*, ed. R.E. Dunlap and R. Brulle, 300–332. New York: Oxford University Press.

McGann, James. 2016. 2015 Global Go to Think Tank Index Report. *Think Tanks and Civil Societies Program (TTCSP)*, September 2. http://repository.upenn.edu/cgi/viewcontent.cgi?article=1009&context=think_tanks

———. 2017. 2016 Global Go to Think Tank Index Report. *Think Tanks and Civil Societies Program (TTCSP)*, January 26. https://repository.upenn.edu/cgi/viewcontent.cgi?article=1011&context=think_tanks

McGann, J.G. 2010. *The Global Go-to Think Tanks*. Philadelphia: Pennsylvania University, The Think Tanks & Civil Society Program.

McGann, J.G., and E.C. Johnson. 2005. *Comparative Think Tanks, Politics and Public Policy*. Cheltenham/Northampton: Edward Elgar.

McGann, J.G., and R. Sabatini. 2011. *Global Think Tanks: Policy Networks and Governance*. London: Routledge.

McGann, J.G., and R.K. Weaver. 2000. *Think Tanks and Civil Societies. Catalysts for Ideas and Action*. New Brunswick/London: Transaction Publishers.

Medvetz, Thomas. 2012. *Think Tanks in America*. Chicago/London: Chicago University Press.

Mirowski, P. 2013. *Never Let a Serious Crisis Go to Waste: How Neoliberalism Survived the Financial Meltdown*. London: Verso.

Osborne, T. 2004. On Mediators: Intellectuals and the Ideas Trade in the Knowledge Society. *Economy and Society* 33 (4): 430–447.

Pielke, Roger. 2007. *The Honest Broker. Making Sense of Science in Policy and Politics*. Cambridge: Cambridge University Press.

Ricci, D. 1993. *The Transformation of American Politics: The New Washington and the Rise of Think Tanks*. New Haven: Yale University Press.

Rich, Andrew. 2004. *Think Tanks, Public Policy, and the Politics of Expertise*. Cambridge: Cambridge University Press.

Ruser, Alexander. 2013. Environmental Think Tanks in Japan and South Korea: Trailblazers or Vicarious Agents? In *Nature, Environment and Culture in East Asia. The Challenge of Climate Change, Climate and Culture Series*, ed. Carmen Meinert, vol. 1, 319–351. Leiden: Brill.

Snider, J.H. 2009. Strengthen Think Tank Accountability. *Politico*, March 2.

Stone, D. 1996. *Capturing the Political Imagination. Think Tanks and Public Policy*. London: Frank Cass.

———. 2004. Introduction. In *Think Tank Traditions: Policy Research and the Politics of Ideas*, ed. D. Stone and A. Denham. Manchester: Manchester University Press.

Stone, D., and A. Denham. 2004. *Think Tank Traditions: Policy Analysis Across Nations*. Manchester: Manchester University Press.

Sheingate, A. 2016. *Building a Business of Politics: The Rise of Political Consultancies and the Transformation of American Democracy*. Oxford: Oxford University Press.

Wagener, Volker. 2016. Angela Merkel's CDU and the Green Party: The Long Road to a New Alliance. *DW*, May 2. http://www.dw.com/en/angela-merkels-cdu-and-the-green-party-the-long-road-to-a-new-alliance/a-19229817. Accessed 29 Aug 2017.

Weaver, Kent R. 1989. The Changing World of Think Tanks. *PS: Political Science & Politics* 22: 563–578.

Weidenbaum, M. 2010. Measuring the Influence of Think Tanks. *Social Science and Public Policy* 47: 134–137.

Weissbach, Michael. 1984. Green Party Gains Threaten Kohl Regime. *Executive Intelligence Review* 11 (40): 32.

Wiedemann, Gregor, Matthias Lemke, and Andreas Niekler. 2013. Postdemokrate und Neoliberalismus – Zur Nutzung neoliberaler Argumentationen in der Bundesrepublik Deutschland 1949–2011. *Zeitschrift für politische Theorie Jg* 4 (1): 80–96.

Winheller, Stefan. 2010. Nonprofit-Organisationen in Deutschlnad un in den USA. *Zeitschrift für Stiftungs- und Vereinswesen* 3: 81–90.

Zuckerman, Alan S. 2005. *The Social Logic of Politics. Personal Networks as Context for Political Behaviour*. Philadelphia: Temple University Press.

CHAPTER 4

Heated Debates and Cooler Heads: Think Tanks and Climate Politics in the United States

Introduction

> What else can happen when men use science and every new thing that science gives, and all their available intelligence and education to manufacture wealth and appliances, and leave government and education to the rustling traditions of hundreds of years ago? (H. G. Wells 1914)

Writing on the eve of World War I, H.G. Wells expressed the fears and hopes of his time. Major scientific and technological breakthroughs had fundamentally transformed everyday live. Cars slowly but inexorably replaced horse-drawn carriages, electricity provided safe and cheap energy, and the invention of the telephone allowed long-distance communication in real time. Nineteenth-century inventions had changed the workplaces and cities, improved the flow of information, and enhanced the mobility of people: Science had, in an impressive way, proved itself to be the major driving force of technological *and* social change.

It is no coincidence that the first think tanks emerged at the same time. Thinks tanks were born out of the optimism and the fears of people who had witnessed the transformative power of scientific thought.

Think tanks in the US were founded to carry the hopes of optimistic philanthropists who trusted the social sciences to bring "social progress" much like the natural sciences were driving technological advancement. Early think tanks were seen as important facilitators of the modernization of policymaking and public life. In order to keep pace with the accelerated

A. Ruser, *Climate Politics and the Impact of Think Tanks*,
https://doi.org/10.1007/978-3-319-75750-6_4

and more complex modern life enabled by natural science knowledge, social science expertise had to be made available to policymakers and administrators to keep pace with rapid social change.

In the past 100 years of their existence, think tanks have changed. So has their image and the expectations of their services.

Investigating the evolution of think tanks in the United States and the increasing differentiation of the think tank landscape is important not only to understand the historical trajectories of their development but also to comprehend the complex interplay between developments at the organizational level and the wider, intellectual environment. As various scholars have pointed out (e.g. Abelson 2004; Rich 2004) the evolutionary development, the numerical growth and important changes in the organizational structure of think tanks can be directly related to a changing political and even intellectual environment. Think tanks inhabit a specific institutional "habitat" and echo a certain intellectual and political climate.

From Enlightened Advisors to Partisan Issue Advocates: The Evolution of Think Tanks in the United States

Think tanks in the United States first emerged around the turn of the twentieth century, a period referred to as "Progressive Era" (Rich 2004: 34). Science in general and the emerging social science in particular were believed to being capable of informing political and social problem-solving, hence contributing to a "rationalization" of political decision-making. The conviction that the social and economic conditions could be substantially improved by the application of social science knowledge encouraged the foundation of institutions capable of providing such invaluable scientific advice. Early think tanks, like the Russell Sage Foundation, which came into being as early as 1907, indicated a specific form of "scientizing" philanthropy and charity. Accordingly, one shouldn't necessarily think of Progressive Era think tanks as "disinterested" (Merton 1973: 275) research organizations adhering solely to the standards of good scientific practice. They were expected to provide "systematic solutions for the broader social problem" (Rich 2004: 34). However, it was a profound trust in the social sciences' ability to solve social problems, which turned early think tanks into frontrunners for "depoliticizing public decision making" (ibid.: 35). This optimistic view of the problem-solving capacities of applied social sciences was further supported by the dominant "linear" understanding of the science-politics relation (see Chap. 3 and Oreskes and Conway 2008: 79): Scientific research was believed to be crucial for framing problems and formulating solutions. Science was thought to immediately compel action by providing decision-makers with *rational instructions* how to solve a given problem.

Early think tanks like the Russell Sage Foundation, the Carnegie Endowment for International Peace (founded in 1910), the Brookings Institution (founded in 1919), and the Hoover Institution (founded in the same year) were set up and expected to '[d]eveloping their own areas of expertise in an environment insulated from the partisan interests of board members and from the vicissitudes of American politics' (Abelson 2004: 217). Think tanks in the early twentieth century benefitted from this general belief in the capabilities of (social) scientific expertise in two different ways: The authority of scientific knowledge itself allowed the Carnegie and Russell Sage Foundations, Brookings, or the Hoover Institution to develop their research agendas without having to fear political interference. Following a technocratic understanding of the science-politics relation, these institutions were trusted to produce state-of-the-art knowledge and eventually provide the best possible advice.

Moreover, the optimisms of the Progressive Era ensured a steady stream of financial resources to these early think tanks. Endowments and donations by wealthy philanthropists and foundations convinced that academic research is key in pursuing their charitable goals, made think tanks 'less vulnerable to the partisan pressure' (Abelson 2004: 218), and allowed them to develop into organizations which resembled very much "universities without students" (Weaver 1989: 566). But the Russell Sage Foundation, Brookings, and others didn't only adhere to the standards of good scientific practice. They also emulated the strategies of universities to exercise influence: 'They hold conferences, seminars and workshops, maintain close ties to the academic community and require scholars to publish articles and books' writes Abelson (2004: 218) pointing out that the think tanks of that era thought of themselves and have to be thought of as academic organizations focusing on *applied* social sciences.

Science and the War Effort: World War II and the Advent of "Big Science"

Changes in the organizational structure and the purpose and the target audience of think tanks occurred not because social sciences proofed to be *less* but *more* effective to solve social problems than initially thought.

The true value of science became visible during the World War II. It was scientific research that led to the technological progress necessary for modern warfare culminating in the Manhattan Project and the invention of the atomic bomb. It had been these close affiliations between scientists (in particular, physicists and mathematicians) and the government, which nurtured an understanding of science serving political goals. Equally important but perhaps more surprisingly, scientists were able to contribute to solving non-technological problems, thus fostering the belief that social, organizational, and political problems could be "managed".

As Fortun and Schweber (1993) have aptly demonstrated, "operation research" units were particularly important for establishing a favourable environment for "applied (social) sciences". The term "operation research" was first established by the British military which was searching for ideas for the efficient use of a recent invention: The radar, 'Roughly a half dozen scientists were involved: they were concerned with the problems of the location of radar sets, the interpretation of radar signals and the efficiency of operation' (Fortun and Schweber 1993: 601).

The main challenge was not so much technical but psychological. To use the radar efficiently in anti-aircraft operations, one must not only detect an enemy aircraft but predict its course, that is, to anticipate the decisions of the pilot. The problem was solved by applying a simple, established economic theory: the rational actor model. Although the theory failed to catch the complexity of human decision-making, it was good enough to operate anti-aircraft guns with sufficient accuracy. Apparently successful, operation research was soon adopted by military planners in the US. Moreover, the successes of operation research proofed the significance (and the prospects) of linking technical aspects of a given problem with organizational, psychological, economic, or sociological questions.

Although early operation research groups had been dominated by physicists with 'psychologists and social scientists appear to have been missing' (Trefethen 1954 cited after Mirowski 2002: 185), the topics and problems studied were closely related to or included social science research. Lacking the nominal expertise in the social sciences of their time but facing the challenges to include human behaviour to solve the problems, they were intended to solve (e.g. military tactics) members of OR teams employed methods and tools familiar to their disciplines (physics and mathematics). Questions of military strategy and tactical challenges were reformulated in the terminology of "game theory" (Mirowski 2002: 186) to arrive at "calculable" solutions.

Think tanks with close ties to the military were among the first to adopt the organizational and methodological insights of the operation research units.

Seeing the 'need for retaining the service of scientists for government and military activities after war's end' (Fisher and Walker 1994: 1) motivated military and political planners to set up the "project RAND" under contract to the Douglas Aircraft Company (ibid.). The RAND Corporation can therefore be seen as an instalment of the war-time operation research units. Although the high hopes of the Progressive Era had been diminished (Rich 2004: 41), scientific research and expertise were still regarded as key for the solution of social problems. However, increasingly deviating from technocratic positions, science was expected to provide a service to decision-makers. Much like in operation research units, authority remains with decision-makers in the military and political bodies. The intellectual and political climate after World War II favoured a new type of think tanks: Government contract research organizations. These new organizations were still applying state-of-the-art scientific theories, methods, and models (cf. Fisher and Walker 1994: 2–3). Government contract research institutions (GCRIs) such as the RAND Corporation differed from earlier academic think tanks not in what they did, but rather in how they related to public authorities. Over time, GCRIs developed close ties with their clients, simultaneously increasing their reliability and dependency on government funding (Abelson 2004: 220).

Experts with a Mission: Advocacy Think Tanks and the New Politics of Knowledge

By moving closer to political and military authorities, GCRIs paved the way for a new form of *advocacy* think tanks. As outlined in the previous chapter advocacy think tanks have to be seen as a major "evolutionary step" in the development of think tanks. In contrast, academic think tanks and GCRIs whose research *agenda* might be influenced by wealthy donors and/or influential clients, advocacy think tanks tailor the *outcomes* of their "research" to the needs of their clients. Advocacy think tanks have a mission. Their primary (if not only) goal is to provide evidence for a particular political, normative position and to disprove and attack dissenting views. Accordingly, it comes as no surprise that advocacy think tanks are frequently accused of "abusing" science (Oreskes and Conway 2008: 60) and being ideological crusader rather than proper research organizations (Klein 2014: 39). However, to understand why there was a '[b]reaking with the tradition established by Robert Brookings, Andrew Carnegie and founders of other early twentieth-century think tanks who were determined to insulate their scholars from partisan politics' (Abelson 2004: 218), one has to consider that the GCRIs had already closed the gap between science and politics by adjusting their respective research agendas to the needs and expectation of their clients. Moreover, by the late 1970s and early 1980s, the linear model of the science-politics relation had ceased to be the dominant concept. Policymakers of this time realized that even the most "accurate" scientific findings cannot compel immediate action. The need for a translation of scientific research and the possibility of interpret research to provide clients political room for manoeuvre changed the relation of think tanks and their clients. Ties between contracted research organizations and clients were based on a shared believe system and normative convictions that determined how scientific information would be interpreted.

The George C. Marshall Institute established in 1984 is a textbook example for this new type of academic think tanks which turned their backs on the scientific community and focused on influencing policymaking processes by affecting the way a political problem is framed and discussed. With the acceptance of fellow researcher becoming less relevant, advocacy think tanks focus on other target audiences. Edwin Feulner, former and current president of one of the most influential advocacy think tanks, the conservative Heritage Foundation, frankly admitted that 'our role is trying to influence the Washington public policy community' (Abelson 2004: 220). He further added some information on how this could be achieved: In contrast to most academic think tanks and GCRIs, advocacy think tanks explicitly address national media outlets to enhance the reach of their respective message (ibid.). Feulner's explanations on the mission, the strategies, and the target audience of advocacy think tanks such as the Georg C. Marshall Institute and the Heritage Foundation indicate important changes in the organizational structure of these new types of think tanks. In more traditional, academic think tanks, scientists were the most important staff members. Advocacy think tanks, however, shift the emphasis from producing knowledge to communicating their preferred message. Accordingly, securing access to political decision-makers and the media became more and more important. As Donald Abelson has pointed out, securing, maintaining, and expanding access is key strategic goal of Edwin Feulner's Heritage Foundation: 'In 2007 Heritage spend close to $8.3 Million, or 17 percent of its $48million budget on media and government relations' (Abelson 2009: 86).

In order to serve as "one-stop policy shops" (Abelson 2009: 88), advocacy think tanks adjusted their publication strategy to the need of political decision-makers and media representatives. For instance, Abelson found that "between 1998–2008, well over 1000 articles written by Heritage scholars appeared in some of America's leading newspapers, including the *Chicago Tribune*, the *Christian Science Monitor*, the *Los Angeles Times*, the *New York Times*, *USA Today*, the *Wall Street Journal*, the *Washington Times*, and the *Washington Post*"(Abelson 2009: 87).

Expertise provided by advocacy think tanks resonates particularly well with journalists, speech writers, and policymakers because it is deliberately tailored to their needs: Op-eds, commentaries, concise formulations, and sound bites as handy, ready-to-use, and free-of- charge information are in high demand with journalist and political advisors alike. Inviting think tank staffers to write guest commentaries, serve on expert panels, or appear as guests in talk shows was and is attractive to media professionals and the political elites because these new advocacy think tanks were able to comment 'on a range of domestic and foreign policy issues' (Abelson 2009:88). In addition, they 'appeal to journalist who are consciously looking for a particular political perspective on an issue' (ibid.)

Are these developments indeed indicating a "breaking with the traditions" established by academic think tanks and GCRIs as proposed by Donald Abelson (2004: 220)? Or should advocacy think tanks be regarded as an evolutionary step as Andrew Rich suggests (2004: 49)? Displaying some similarities with GCRIs, advocacy think tanks are equally dependent on clients who are willing to pay and, even more importantly, to listen to their message.

The "breach" however cannot be explained without considering the fundamental changes to the political culture. The "evolutionary step" to advocacy think tanks can be interpreted as an adaption to the increasing polarization of the political and media landscape in the United States. In short, if think tanks had changed over time, so had their clients.

Expertise for a Divided Political Landscape

The end of the Progressive Era which gave way to a more instrumental understanding of scientific knowledge was not the only change in the environment that affected the development of think tanks in the US. Political scientists provide evidence for an increasing polarization of the American party system that can be traced back to the late 1970s: 'The two-party system in the US does not always lend itself to compromise and bipartisan negotiation, but this has been particularly true in the past few decades. Empirically, polarization among party elites has risen dramatically' (Kuo and McCarty 2015: 49).

The increasing polarization is, as Kuo and McCarty continue to explain, not resulting from centrifugal powers pushing Republicans and Democrats to the extremes. They rather state that '[t]he changes are driven primarily by a marked movement to the right by the Republicans' (ibid.: 50). This development within the Republican Party has been attributed to the more recent successes of the "Tea Party Movement" and the "Freedom Caucus" in the House of Representatives:

> A small set of nationally operating Republican elites, many of whom have been promoting a low-tax, anti-regulation agenda since the 1970s, have played a key role in local and regional Tea Party efforts. These elites have long since developed a policymaking infrastructure in Washington, but had previously achieved only limited success in directly connecting themselves to an activist grassroots base. (Williamson et al. 2011: 26)

Resulting from this development was the spread of an increasingly uncompromising approach to policymaking and a rejection of bipartisan solution (Ruser and Machin 2017; Williamson et al. 2011: 36). Moreover, as, for instance, Thomas Frank has argued, the political divide was accompanied by a wider cultural movement or "backlash" (Frank 2004: 20ff) which culminates in the talk of the "two Americas" (ibid.: 13). According to the narrative of the two Americas, the US is divided into two irreconcilably confronted groups of red-state, or "Heartland" Americans living in rural America (sometimes further narrowed down to the so-called rustbelt states) on the one side, and a liberal elite residing conveniently and ignorant of the concerns and feelings of "true Americans" in the cities on the coasts. Although simplified the narrative of the two Americas provided policymakers, journalists, grassroots activists, and conservative voters with a comprehensive framework for how to position themselves towards a given political, economic, and normative challenge.

It is this simplified depiction of the political and social cleavages in America that allowed Rand Paul to reject climate politics as "anti-American and anti-freedom" (Antonio and Brulle 2011: 195). And, it is this polarizing narrative which played an important role in rise of the advocacy think tanks:

> Between 1970 and 1996, the number of think tanks operating in the United States grew from fewer than 60 to more than 300 (...). This proliferation of new think tanks occurred during a period when new interest groups and other types of political organizations were also forming in great number in the U.S., with a common eye toward contributing to and influencing public policy debates. (Rich 2001: 54)

The rise of advocacy think tanks can be seen as a demand-driven reaction to an increasing polarization of the political landscape. The founding of advocacy think tanks fits well into a political system which witnessed related developments such as gerrymandering and an increasing polarization in the media landscape.

Gerrymandering, that is, the deliberate drawing of the borders of electoral districts to get a political advantage, had an immense impact on American politics:

> The division of the United States into predictably "red" and "blue" states and the gradual decline in the number of genuine "battlegrounds" (...) where either party has a genuine chance of victory suggests that American politics today is more polarized than in eras past. (Iyengar and Hahn 2009: 19)

In consequence, serving preferences (and prejudices) of carefully defined electorates became more important while at the same time rendering the ability to compromise less important. This development explains why climate politics could become an exceptional prominent battleground for advocacy think tanks:

> Because it is hard to understand climate change from personal experience, people often rely on others presumed to be more expert to answer their questions about climate change. For the most part, they do not go to climate scientists directly but rather to intermediary sources, pre- dominantly

in the mass media, that present information and opinions in language and graphics that are easy to comprehend (Weber and Stern 2011: 320).

Communicating climate change to the wider public follows the rule and norms of journalism. Yet because of the complexity of the issue and the opaqueness of scientific language, the writing on climate change and climate politics is particularly demanding for journalist, in turn, explaining the demand for comprehensible, easy-to-use expertise provided (among others) by think tanks. Yet, while meeting the normative and practical standards of good journalism might be particularly hard, some norms can actually contribute to a misrepresentation of the complex issue of climate change.

As Maxwell and Jules Boykoff have pointed out, one of the most important rules is the journalistic norm of balance (Boykoff and Boykoff 2004: 126). Intended to guarantee the "neutrality" of media coverage by demanding to representing "both sides" of a story, balanced reporting runs into problems when neither the credibility nor the structure or the size of the two sides can be estimated: 'In fact, when it comes to coverage of global warming, balanced reporting can actually be a form of informational bias' (ibid.).

The problem is to decide when dissenting views have to be presented as the "other side" to a story. For instance, comparing the International Panel on Climate Change (IPCC) and the Nongovernmental International Panel on Climate Change (NIPCC) reveals that the IPCC is by far the larger and more prestigious organization. Set up by the Word Meteorological Organization and the United Nations Environmental Programme in 1988, the IPCC summons hundreds of the most renowned climate scientists and issues comprehensive reports on the state of climate research. The NIPCC on the other hand was founded by Fred Singer[1] and is financed by the Heartland Institute, an advocacy think tank known for its climate sceptic views (Dunlap and McCright 2011: 149) which is also a member of the "Cooler Heads Coalition" (see below). Does a balanced coverage require journalist to cite the NIPCC report *Why Scientist Disagree About Global Warming*[2] alongside the latest assessment report of the IPCC? To draw a distinction between the credibility of the IPCC and the NIPCC requires *external* criteria (e.g. scientific quality criteria or academic reputation). However, since external criteria can be difficult to define and might be contested.

As described in the Chap. 2, the credibility and impartiality of the IPCC was seriously challenged in 2009/2010 when computer hacker broke into the server of the University of East Anglia. Data and email correspondence released in the wake of the hacker attack created the impression that leading climate scientists such as then director of the Climate Research Unit (CRU) Phil. D. Jones were deliberately refusing to share data and give biased information. The incident which was soon referred to as *Climategate* by climate sceptics (in particular in the USA) triggered thorough investigation and stimulated further debate on the scientific practices and transparency. Although a government commission found that '[t]he evidence that we have seen does not suggest that Professor Jones was trying to subvert the peer review process' (Government Response to the House of Commons Science and Technology Committee Report *The disclosure of climate data from the Climatic Research Unit at the University of East Anglia*, 2010), the "climategate" incident was a welcome occasion for climate sceptics to question the credibility of the IPCC and reaffirming claims of a biased academic science.[3]

The most important consequence of a (alleged) scandal-like climategate is that it provides journalists with incentives to invite "the other side" to tell its story. With (parts of) their audience in doubt on the credibility of professional climate scientists refusing to report dissenting views could be interpreted as biased reporting.

Journalists have to respond to their respective audience. When audiences change, for example, as a consequence of an increasing polarization of the political landscape, media reporting will change too:

> As media audiences fragment, television networks and programs now cater to specific segments of the public rather than to the masses. At the same time, shifting structural, economic, and audience conditions are helping to erode the boundaries between news and entertainment, so that entertainment values now filter into hard news programs and vice versa.
>
> Against this backdrop, cable news outlets have begun to appeal to particular segments of the audience with targeted political messages. This proliferation of opinion and overt partisanship has been plainly observed by popular commentators. (Feldman et al. 2012: 6)

This depiction of the medial landscape in the USA corresponds with and adds to the description of the United States as market-oriented knowledge regime. The 'highly adversarial, partisan, and competitive knowledge production process' is mirrored by a fragmented audience. TV stations, newspaper, and other news outlets specialize to cater to the needs of these isolated audiences.

As Baum and Groeling demonstrate, the process of fragmentation has further been increased by the growing importance of online sources:

> Many (...) Internet outlets—including, but not limited to, blogs—are overtly niche-oriented, seeking to attract a smaller, but more loyal, segment of the overall audience.5 While political partisanship is by no means the only dimension upon which niche-marketing strategies might be based, in the realm of political information, partisan- ship is one of the key lines of demarcation allowing Web sites to attract a relatively loyal audience. (Baum and Groeling 2008: 347)

Getting access to the media should therefore be relatively easy for advocacy think tanks.[4] With agreed upon criteria to exclude viewpoints from the debate missing and a media landscape that increasingly caters to the needs and expectations of a fragmented audience, tendentious expertise is welcomed by journalists who seek to "balance" their reports and media representatives who are looking for like-minded experts. Lauren McDonald finds that 'FOX television was three and a half times more likely to cite conservative think tanks than centrist think tanks on education stories' (2014: 866), and Antonio and Brulle state that '[r]eeling from conservative attacks over liberal bias, "mainstream media," seeking "editorial balance," often grant parity to "climate skeptic" news releases and policy papers, from right-wing think tanks and their bought experts and pundits, with peer-reviewed science' (2011: 197).

Think tanks and advocacy think tanks in particular fit in the political and media landscape. Their services are valued by policymakers and journalists who have to cater to the expectations of a divided audience and polarized electorates.

Expertise for "Anti-American" Climate Politics

> *Global warming isn't just one of many public policy debates that free-market advocates need to win. It is a war, the most important and most consequential war of our era. Progressives have declared war on capitalism and the technologies, fuels, and industries critical to its survival. This is why they took over the environmental movement in the 1980s and 1990s; not to protect the environment, but to wage war on capitalism.* (Bast 2017)

Joseph Bast, CEO of the conservative Heartland Institute, couldn't be more clear. Climate science and climate politics aren't just a controversial political issue. It is a frontline in a war of ideas. A war that involves disagreement about climate science but which is mainly fought to defend the specific variety of American capitalism.

The previous section described political and media landscape in the USA highlighting the increasing polarization and linking it to the demand for the services of (advocacy) think tanks. While this helps understanding why think tanks could develop into important actors in the US, it has yet to be investigated what exactly think tanks are trying to sell on the American marketplace of ideas.

The following paragraphs will deal with the content and the substance of the debate on climate politics and climate science in the United States. It will be demonstrated that climate change isn't depicted as an environmental challenge only and that the rift between "climate believer" and "climate deniers" or sceptics is rooted in an underlying dispute about the role of markets, government, and a particular understanding of individual freedom.

> Despite the growing consensus within the scientific community regarding global warming, the success of the environmental community in getting global warming on the national agenda, and the receptive nature of public opinion, it nevertheless appears that claims about the existence of global warming became more contested in the United States policy arena in the late 1990s-with the result that effective policy-making ground to a halt. (McCright and Dunlap 2003: 349)

Contesting climate science and the political measures derived from its findings and recommendations as described by Aaron McCright and Riley Dunlap are consistent with some aspects of political culture and, if you will, cultural folklore. It is important to note that advocacy think tanks didn't "invent" a general scepticism towards climate science but rather build on and exploit existing, traditional scepticism of experts and centralized planning.

For instance, already in the 1960s, Richard Hofstadter pointed out that the end of the Progressive Era was accompanied by an increasing alienation of the "wider public" from intellectual community (1963: 39), thus reaffirming a deeply rooted "anti-intellectualism".

Linking anti-intellectualism to a conservative attitude, Hofstadter writes that '[i]n the course of generations, those who have suffered from the operations of intellect, or who have feared or resented it, have developed a kind of counter-mythology about what it is and the role it plays in society' (ibid.: 45). This anti-intellectual counter-mythology is

> 'founded upon a set of fictional and wholly abstract antagonism. Intellect is pitted against feeling, on the ground that it is somehow inconsistent with the warm emotion. It is pitted against character, because it is widely believed that intellect stands for mere cleverness which transmutes easily into the sly or the diabolical. It is pitted against practicality, since theory is held to be opposed to practice, and the "purely" theoretical mind is so much disesteemed. It is pitted against democracy, since intellect is felt to be a form of distinction that defies egalitarianism' (Hofstadter 1963: 45–46).

The counter-mythology seems to be at work when "policies to combat global warming" are decried by the Heartland Institute as 'a green Trojan horse, whose belly is full with red Marxist socioeconomic doctrine' and part of a mere conspiracy contrived by climate scientists to enrich themselves and ill-meaning politicians determined to destroying the traditional, humble, American way of life (Klein 2011).

Also, the counter-mythology helps to understand why climate denial 'emanated almost entirely from the right wing of the American political spectrum' (Oreskes and Conway 2008: 59) and why climate science is particularly prone to becoming politicized in the US.

Finally, and most importantly, the counter-mythology helps understanding why and how (conservative) advocacy think tanks play an ambivalent role in the dispute on climate science. For they arguably play a double game: Challenging climate science by providing "counter-evidence" and exploiting anti-intellectual sentiments predominant in conservative circles in the US.

> What I fail to understand is why global warming has come to be viewed as a political or ideological issue (…). If you are in a house where there's a strong burning smell and the air is getting smoky, the sane response is to acknowledge that there is a fire somewhere and do something about it – no matter what one's political ideology might be. (Oreskes and Conway 2008: 59-60)

This letter to the editor of *The New York Times* cited by Oreskes and Conway appeals to the common sense. If you have evidence of something, be it a fire or climate change, forget about political or ideological differences and act accordingly.

While the writers' astonishment of the politicization of climate change politics is understandable, the letter is (unconsciously) sketching out the two most important components of climate denial: (1) Questioning the soundness of the underlying science and (2) invoking "common sense" arguments to disprove the "alarmists" claims.

'Sometimes, it just boils down to common sense – when you consider that an iceberg melts in one part of Greenland while residents in its capital city, Nuuk, feel as if they were entering another Little Ice Age due to a record-cold June. Or, the shipload of climate change scientists which set sail on a "we told you so" mission only to get stuck in the ice in Antarctica during the Southern Hemisphere's summer last year, surrounded by so much ice, two icebreaker rescue vessels couldn't cut through to their aid' writes Susan Brown in June 2015 in *The Christian Post* evocating the (in) famous image of Oklahoma Senator James Inhofe tossing a snowball to the senate floor to disprove global warming.[5]

The two components of the denier's strategies correspond with the two frontlines in the "global warming war". Conservative advocacy think tanks are engaged in providing ammunition for both battlegrounds. They help challenging climate science and contribute to attempts to evoking common sense arguments against climate politics.

Conservative forces are invoking anti-intellectual sentiment (see above) to reject the authority of climate science and climate scientists. Aaron McCright and Riley Dunlap observe that

> 'conservative think tanks and the Republican Party have regularly disparaged mainstream scientists and the pronouncements of the scientific community's most prestigious bodies, while promoting the largely debunked claims of a handful of climate change contrarians (...). This conflict reflects a deeper division between those who levy critiques of the industrial capitalist order and those who defend the economic system from such challenges (...). Our results provide strong evidence that the long-term divide over global warming between elites and organizations on the Left and the Right has in recent years emerged within the general public as well' (2011a: 166).

Investigating this "general public view", McCright and Dunlap (2011b) found that climate change might not only be an existential threat to human well-being but also to the core values and self-image of conservative, predominantly white and male Americans. In an empirical study, they showed that conservative whites were key players in spreading climate sceptic positions (e.g. as conservative talks radio hosts or CEO of conservative think tanks) *and* the primary target group of the climate sceptic message.

Climate change, and in particular comprehensive environmental *regulation*, that is portrayed as a necessity to mitigate its direst consequences, therefore, is a direct threat to a conservative depiction of the "American way". A changing climate can change the "terms of trade" of largely fossil fuel-based economies, and climate politics threaten a lifestyle that is based on individual freedom and unsustainable consumerism. In consequence, it 'is natural for individuals to adopt a posture of extreme skepticism, in particular when charges of societal danger are leveled at activities integral to social roles constructed by their cultural commitments' (Kahan 2007 cited after McCright and Dunlap 2011b: 1171).

It is important to stress that conservative think tanks are not speaking for business interests only—although they might be dependent on their financial support, cf. Mayer 2016)—but also speaking up for conservative groups who seemingly experience climate change and, more importantly, climate politics as an immediate assault on their personal way of life:

White (male) conservatives seem to be particularly responsive to the messages of conservative think tanks and therefore become a likely target group. As, for instance, Naomi Klein (2014: 55–56, 60–62) and Naomi Oreskes and Eric Conway (2010: 208) observe, think tanks pick up and affect an anti-climate sentiment predominant within a group of people for whom '[i]t's rational (…) to deny climate change – to recognize it be intellectually cataclysmic' (Klein 2014: 61). According to McCright and Dunlap, the fight on climate politics is the most doggedly fought by those whose very identity seems to be under attack:

> [T]his pattern—where conservative white males are more confident in their knowledge of climate change than are other adults, even as their beliefs conflict with the scientific consensus—is consistent with our expectation that identity-protective cognition and system-justifying tendencies are especially strong within conservative white males. Such processes, we argue, lead them to reject information from out- groups (e.g., liberals and environmentalists) they see as threatening the economic system, and such tendencies provoke strong emotional and psychic investment, easily translating into (over)- confidence in beliefs. (McCright and Dunlap 2011b: 1167)[6]

For conservatives in the Unites States, climate politics isn't about protecting the environment but is seen as just another form of government regulation which generally 'falls under interference with the pursuit of self-interest by people trying to make a living, people using their self- discipline to become self- reliant' (Lakoff 2002: 211).

To a large degree conservative resistance to climate protection isn't rooted in a diverging interpretation of "facts" or alternative information. Climate politics is an attack on core values and some of the most fundamental moral convictions. Climate protection isn't only constraining the use of natural resources but interfering with one's personal freedom.

Climate politics thus became just another front in an 'ideological war aimed at freeing American business from the grip of government' (Mayer 2016: 120) since 'environmental, worker-safety, and product-safety regulations (...) are too cumbersome and get in the way of business' (Lakoff 2002: 211).

Funding Relationships and Social Capital

In order to become a vital part of conservative policy networks, think tanks have to prove their like-mindedness by adopting the whole free-market ideology rather than focusing on environmental problems and climate politics alone. As Robert Brulle (2014) has shown, conservative think tanks that focus on a wide range of issues were particularly important in starting and maintaining a "climate change counter movement" (CCCM) (2014: 682).

By conducting an income analysis alongside an analysis of the foundation funding Brulle finds that the 'single largest funder are the combined foundations Donors Trust/ Donors Capital Fund. Over the 2003–2010 period, they provided more than $78 million in funding to CCCM organizations' (Brulle 2014: 687). Brulle further shows that 'conservative think tanks were the larges recipients of foundation support' (ibid.) and that the majority of this support (77.4% of the total funding) comes from a relatively small number of funding organization (22) (ibid.: 691). Brulle's findings are consisting with more recent data obtained from "990 forms" ("Return of Organizations Exempt from Income Tax under section 501(c), 527, or 4947(a)(1) or the Internal Revenue Code"). In 2014 four large donors, the Holman Foundation, the Donors Trust, the Donors Capital Fund, and the Charles G Koch Charitable Foundation, accounted for the most important sources of income for conservative think tanks from the Cooler Heads Coalition. The Heartland Institute, for instance, lists a total of incoming grant money of $6.890.995 with $1.900.000 coming from a single donor: the Donors Capital Fund. In the same year, the CATO Institute received $1.144.390 form Charles G Koch Charitable Foundation, $252.800 from the Donors Trust, $100.000 from the

Donors Charitable Fund, and 50.000$ from the Holmes Foundation. Of the $7.105.791 of incoming grants, the Competitive Enterprise Institute received in the same year $1.193.850 came from the Donors Trust, $200.000 from the Donors Capital Fund, and $53.924 from Charles G Koch Charitable Foundation.[7]

'This distribution of funding shows that both conservative foundations and the recipient organizations [predominantly conservative think tanks, AR] are core actors in the larger conservative movement' (Brulle 2014: 688). A subsequent network analysis conducted by Brulle revealed stable, tightly knitted network of conservative think tanks and wealthy funding organizations (Brulle 2014: 690).

The financial success of these think tanks stems from their close ties to wealthy donors who have been engaged in a vigorous fight against big government and for "freeing the markets". As Jane Mayer has pointed out, access to financial resources depends on access to a small group of extraordinarily wealthy and determined actors:

> Fewer than two hundred extraordinarily rich individuals and private foundations accounted for the $750 million pooled by Donors Trust and its sister arm, Donors Capital Fund since 19999. Many were the same billionaires and multimillionaires who formed the Koch network. (Mayer 2016: 347)

The paramount significance of these funding organizations becomes evident when the financial relations are mapped and displayed as a network graph (see Fig. 4.1). According to Borgatti et al. (2009: 894), the social capital of an actor can be determined by locating them within their respective network. As displayed in the figure below, the Heritage Foundation, the AEIPPR, the Heartland Institute, the Competitive Enterprise Institute, the CATO Institute, and the National Center for Policy Analysis had all received substantial single donations from one of the funding sources described above.

The close ties between a relatively small group of (partially interrelated) funding organization and conservative think tanks point at the significance of the context in which conservative think tanks operate.

As displayed in the figure below, some conservative think tanks have access to substantial financial resources. While these economic resources are a prerequisite for the operations, the research, and the media campaigns launched by these organizations, financial leeway alone cannot explain why, for instance, the Competitive Enterprise Institute engages

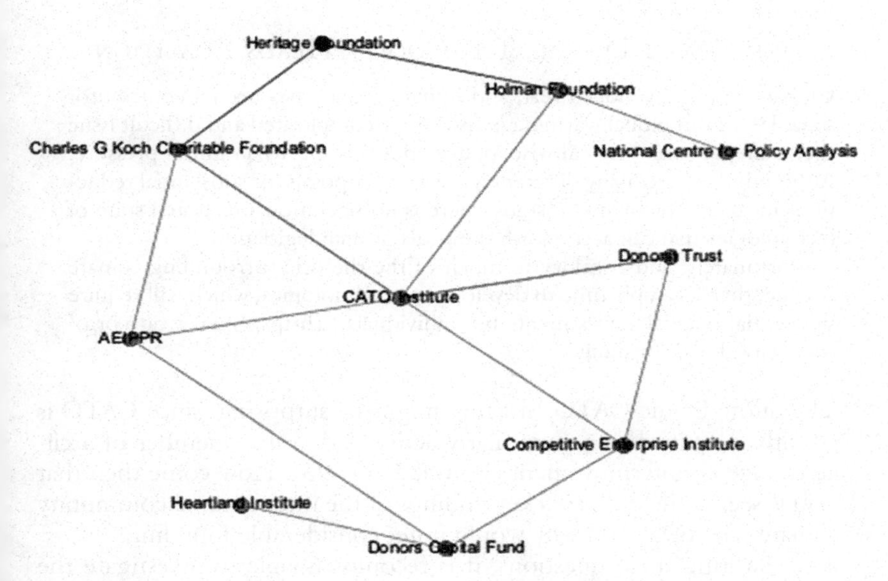

Fig. 4.1 Donor Think Tank Network (year of reference 2014/$30.000 donation). (Source: conservativetransparency.org, own calculations)

the public rather than challenging climate scientists while, for instance, the CATO Institute doesn't seem to be particularly interested in questioning the existence of climate change but focuses on advertising climate engineering.

To understand why conservative think tanks are aiming at specific target audiences and why their strategies include straightforward denial of the scientific basis of climate science (Heartland Institute) alongside an implicit acceptance of climate change (Cato Institute), it is important to investigate the position of these think tanks within larger networks. As will be shown in the following paragraphs, the significance of these networks goes well beyond the distribution of financial resources: The climate change countermovement (Brulle 2014: 681) can be depicted as a network structured by "value homophily" (McPherson et al. 2001: 419). A network based on value homophily would include (only) those actors who share a common set of fundamental beliefs, normative convictions, and political orientations (cf. McPherson et al. 2001: 419, 429), that is, values and beliefs that transcend the more narrow challenge of climate change and climate politics.

Network of Denial: The Cooler Heads Coalition

> Global warming is indeed real, and human activity has been a contributor since 1975. But global warming is also a very complicated and difficult issue that can provoke very unwise policy in response to political pressure. Although there are many different legislative proposals for substantial reductions in carbon dioxide emissions, there is no operational or tested suite of technologies that can accomplish the goals of such legislation.
>
> Fortunately, and contrary to much of the rhetoric surrounding climate change, there is ample time to develop such technologies, which will require substantial capital investment by individuals. (https://www.cato.org/research/global-warming)

The quote by the CATO Institute might be surprising, since CATO is frequently described as a particularly active and visible member of a climate change countermovement (Powell 2011: 93). How come then that CATO is seemingly *accepting* key findings of the international community of climate scientists? And why would it get considerable founding?

To answering these questions, it is recommendable to investigate the context in which conservative think tanks operate in more detail, searching for indicators of value homophily (see above). By analysing member organizations, it becomes possible to focus not only on financial relations but to understand the *anatomy* of "thought collective" (Mirowski and Plehwe 2009) and why it is concerned with climate science and climate politics.

Moreover, considering how such a network "fits in" the wider context of American political debate allows for understanding the specific role think tanks such as the Competitive Enterprise Institute play, which audiences they address, and why they ally themselves with free-market hardliners like the "Americans for Tax Reform".

> The pro-treaty side would have us believe that the underlying science is clear, that the consensus of scientists has spoken, and all the rest of us need to do is get on with saving the planet. But in science, [the discussion] isn't over until a hypothesis has been validated, either by an experimental test or by observations about the real world. (https://cei.org/content/first-cooler-heads-lecture-global-warming)

The statement comes from Marlo Lewis, the vice president for policy and coalitions at the Competitive Enterprise Institute. Mr. Lewis made a speech at the *First Cooler Heads Coalition on Global Warming* in May 1998. Even more interesting than seeing him using the tried-and-tested strategy of demanding more robust evidence of climate change is perhaps the existence of an active coalition of climate sceptics in the year after the Kyoto summit.

The coalition claims to be aiming at a more balanced and unagitated debate. The main slogan used by the coalition is “may cooler heads prevail”, thus calling for “common sense solutions” and warning of hasty conclusions and premature action (cf. Boykoff and Olson 2013: 281). The Cooler Heads Coalition (CHC) is a good example for the ballet of monetary interests, ideological networks, and political cliques that dominate policymaking in Washington.

Although the CHC has seen some changes in its membership, its mainstays consist of established conservative organizations such as the “Americans for Tax Reform”, one of the most vocal advocacy groups focusing on limiting the burden on taxpayers; the “Committee for a Constructive Tomorrow”, a non-profit organization which has promoted free-market solutions for decades; and some of the most visible climate sceptic think tanks such as the “Competitive Enterprise Institute” and the “Heartland Institute” (Table 4.1).

The Cooler Heads Coalition is of particular interest since it epitomizes the increasing polarization of beliefs on anthropogenic climate change in the USA. As Riley Dunlap and Peter Jacques have pointed out (2013: 700), the CHC was set up to replace the Global Climate Coalition (GCC) after its “demise in 2002” (ibid.).[8] The CHC consists mainly of conservative think tanks with the Competitive Enterprise Institute playing the leading role (Dunlap and McCright 2011: 151). Since the majority of the advocacy think tanks that involved not exclusively focusing on climate change but are proponents of limited government and free-market position, they were able to attract considerable funding from wealthy donors (Dunlap and McCright 2011: 147) especially from industries which didn’t want to align with the defectors of CHC’s predecessor, the Global Climate Coalition. The CHC is therefore a good example for how conservative think tanks are not only thriving because of an increasing political divide but also benefit from friction within business elites (cf. Hein and Jenkins 2017).

Moreover, 11 of the past and current member organizations of the coalition were set up between 1982 and 1986, that is, at the height of the “Reagan Revolution” when “free enterprise” was declared “fundamental

Table 4.1 Membership of Cooler Heads Coalition (X = Membership)

Name	*2004*	*2007*	*2008*	*2010*	*2015*	*2017*	*Description*
60 Plus Association	X	X	X	X	X	X	Founded in 1992, promotes limited government and free-market politics
Alexis de Tocqueville Institution	X	X	X	X		X	Founded in 1985, dissolved in 2007
America's Future Foundation			X	X		X	Founded in 1995, network of "liberty-minded young professionals" to promote free trade policy and "liberty" (lobby against regulatory dimension of climate politics)
American Conservative Union					X		Founded in 1964 as "an umbrella organization harnessing the collective strength of conservative organizations fighting for Americans who are concerned with liberty, personal responsibility, traditional values, and strong national defence"
American Energy Alliance					X		Founded in 2008, as an independent grassroots affiliate of the Institute for Energy Research (IER)
American Legislative Exchange Council	X	X	X	X	X	X	Founded in 1973, platform for state legislators and private/business representatives to develop "model bills" on tax reform, budgets, education, and international trade (including climate change)
American Policy Center	X	X	X	X	X	X	Founded in 1988, grassroots action and education foundation to promote free-market policy Publications include the "DeWeese Report" and occasional special reports

(continued)

Table 4.1 (continued)

Name	*2004*	*2007*	*2008*	*2010*	*2015*	*2017*	*Description*
Americans for Prosperity				X	X	X	Founded in 2004, support for "grassroots" organizations (e.g. the Tea Party), campaigning
Americans for Tax Reform	X	X	X	X	X	X	Founded in 1985, "American Taxpayer Pledge"/ Policymakers[a]
Association of Concerned Taxpayers	X						Founded in 1975, promotes limited government and free-market politics
Center for Security Policy	X						Founded in 1988 to promote national security and energy sovereignty
Citizens for a Sound Economy	X						Founded in 1984, dissolved in 2004
Committee for a Constructive Tomorrow	X	X	X	X	X	X	Founded in 1985 "to promote a much-needed, positive alternative voice on issues of environment and development" and seeks to influence public interest debates on environmental issues
Competitive Enterprise Institute	X	X	X	X	X	X	Founded in 1984, CEI is "dedicated to advancing the principles of limited government, free enterprise, and individual liberty" and to promote "more rational" climate politics
Consumer Alert	X						Founded in 1977, dissolved in 2006
Council for Citizens Against Government Waste					X		Founded in 1983, protects "taxpayer dollars", issues reports/brief that attack subsidies for renewables
Defenders of Property Rights	X						Founded in 1991, promotes limited government, opposes environmental regulation

(*continued*)

Table 4.1 (continued)

Name	*2004*	*2007*	*2008*	*2010*	*2015*	*2017*	*Description*
Eagle Forum					X		Founded in 1972, large conservative interest group/ network, "pro-family" forum, actively lobbies against climate politics
Fraser Institute, Canada	X	X	X	X	X	X	Founded in 1974, research on a variety of topics including education, economic freedom, health and labour politics, and climate change, comprehensive reports and briefs emphasize the economic implications of uncertainty in climate models
Freedom Action					X		Subsidiary of the Competitive Enterprise Institute
FreedomWorks		X	X	X	X	X	Founded in 2004, grassroots service centre supporting activists in their fight for lower taxes and less government and more individual freedom, criticizes the regulatory dimension of climate politics, issues "house" and "senate" scorecards to track whether elected representatives vote with FreedomWorks principles
Frontiers of Freedom	X	X	X	X	X	X	Founded in 1995, an "educational foundation whose mission is to promote the principles of individual freedom, peace through strength, limited government, free enterprise, free markets, and traditional American values" by making the positions of "candidates for political office more transparent"

(*continued*)

Table 4.1 (continued)

Name	*2004*	*2007*	*2008*	*2010*	*2015*	*2017*	*Description*
George C. Marshall Institute	X	X	X	X	X		Founded in 1984, dissolved in 2015, focused on impact of scientific knowledge on public policy (e.g. defence and climate politics)
Heartland Institute	X	X	X	X	X	X	Founded in 1984, conservative think tank, which promotes free-market solutions, organizes the climate sceptic Nongovernmental International Panel on Climate Change (NIPCC)
Heritage Foundation	X						Founded in 1973, promotes free-market politics, limited government, individual freedom; opposes environmental regulation
Independent Institute	X	X	X	X	X	X	Founded in 1986, maintains seven issue-oriented research centres. The centre on health and the environment provides scientific evidence to tackle mainstream climate science[b]
Istituto Bruno Leoni, Italy	X	X	X	X	X	X	Founded in 2003, publication of books, Leoni award, Leoni lecture and Leoni index in liberalization; opposition to regulatory aspects of climate politics
John Locke Foundation					X		Founded in 1990, promotes free-market politics, challenges climate science, and opposes environmental regulation
JunkScience.com	X	X	X	X	X	X	Founded in 1996, online portal dedicates to debunk "junk science" including mainstream climate science
Lavoisier Group, Australia		X	X	X	X	X	Founded in 2000, provides collection of climate sceptic papers and news article

(*continued*)

Table 4.1 (continued)

Name	*2004*	*2007*	*2008*	*2010*	*2015*	*2017*	*Description*
Liberty Institute, India		X	X	X	X	X	Founded in 1996, independent think tank that issues reports and books on individual rights, rule of law, limited government, and free markets
National Center for Policy Analysis	X	X	X	X	X	X	Founded in 1983, dissolved in 2017, opposes regulation on healthcare, taxes, welfare, education, and environmental politics
National Center for Public Policy Research	X	X	X		X		Founded in 1982, as "communication and research foundation" to promote strong defence politics and free-market solutions; opposes environmental regulation
Pacific Research Institute	X	X	X	X	X	X	Founded in 1979, operates five research centres including the "centre for environment; promoting free market solutions/fighting environmental regulation"
Seniors Coalition	X	X	X	X		X	Founded in 1990 as a public advocacy group to fight Medicare healthcare plans
Small Business and Entrepreneurship Council		X	X	X	X	X	Founded in 1994 to research and advocate for small business, free entrepreneurship; opposes environmental regulation
Small Business Survival Committee	X						Opposes environmental regulation that could harm small business
TaxPayers' Alliance, UK					X		Founded in 2004 to promote "transparency" and limited government and to "stop the [green] energy swindle"

(*continued*)

Table 4.1 (continued)

Name	*2004*	*2007*	*2008*	*2010*	*2015*	*2017*	*Description*
The Advancement of Sound Science Coalition	X						Founded in 1993 by Philip Morris to challenge "mainstream science" including climate science

[a]The Americans for Tax Reform (ATR) successfully urged conservative policymakers into signing the following pledge:

I, ______________________, pledge to the taxpayers of the state of ______________________, and to the American people that I will:

ONE, oppose any and all efforts to increase the marginal income tax rates for individuals and/or businesses; and

TWO, oppose any net reduction or elimination of deductions and credits, unless matched dollar for dollar by further reducing tax rates. (Americans for Tax Reform, "Taxpayers Pledge")

By tracking whether the respective politician sticks to his or her pledge the ATR is very successful in promoting free-market ideas. (cf. Hacker and Pierson 2005a: 40)

[b]For instance, in a paper published in *The Independent Review*, William Butos and Thomas McQuade state a "boom" in climate science comparing the recent increase in publications on climate change to historical examples of eugenics, which after receiving generous funding from the Nazis, all but vanished after World War II (cf. Butos and McQuade 2015: 166)

Source: https://www.desmogblog.com/cooler-heads-coalition, additional information obtained from websites and mission statements of the respective organization

to the American way of life" while "government should be limited" (Republican Party Platform of 1984, Preamble).

The spirit of free-market politics and attempts to trimming back government influence referred to as "Reagonomics" (see Rousseas 2016) prevailed in the strategies of the Cooler Heads Coalition. For instance, in a briefing at the United States Congress on CO_2 in 2001, sponsored by the Cooler Heads Coalition, Ross McKitrick concludes that the 'optimal policy is to leave CO_2 emissions unpriced and unregulated' (McKitrick 2001: 12). According to the briefing regulating carbon emissions, any regulation will lead to "command-and-control" economics and, eventually, to central planning and a "Carbon Cartel" (ibid.: 2).

No Country for Climate Politics? A Paradise for Conservative Think Tanks?

The United States seem to be a particular fertile soil for conservative, climate sceptic think tanks for various reasons. First, an increasingly polarized political and media landscape creates a growing demand for partisan expertise and advocacy. A network of funding organization supplies like-minded think tanks with substantial resources. The term “like-minded think tank” refers to organizations that share a set of core values that circle around a fundamental belief of the superiority of free-market politics and the primacy of a radical understanding of individual and entrepreneurial freedom. Recurring to this shared belief system helps explaining why conservative think tanks, although launching attacks on “mainstream science”, ‘deal with global warming (…) not simply as economic threats but as threats to free enterprise, individual rights and Western progress’ and carrying out climate change denial ‘with a vociferousness that goes beyond the findings of the IPCC and mainstream climate scientists’ (Dunlap and McCright 2010: 252).

So, what is it then that makes conservative think tanks in the United States successful? And what exactly does “success” means in influencing climate politics?

As was argued in this chapter, think tanks that challenge climate politics are members of a dense network which does not only provide them with substantial financial resources but which build around value homophily. This network of like-minded organizations not only maintains close ties to wealthy donors but is also part of what Jacob Hacker and Paul Pierson call a “new power structure” (2005b: 141). Describing the anatomy of this conservative network, Hacker and Pierson distinguish between “true grassroots” ranging from small, local movements to powerful lobby organizations like the National Rifle Association to elite interest groups, action committees, and wealthy donors. Conservative think tanks, in the view of Hacker and Pierson, are the permanent representatives of this network in Washington. They’re therefore an important cog in the “Republican Machine” (ibid. 135). American politics in general and conservative politics in particular are depicted as “being driven from the top” (ibid.). And this “top” is populated by “new power brokers” within the Republican Party, that is, policymakers who are not only responsive to the conservative, free-market, limited government rhetoric conservative foundations, grassroots movements, lobbyists, and think tanks but who often actively promoting “extreme” views. Moreover, according to Hacker and Pierson, ‘they are powerful not primarily because they hold positions of formal authority but because of their strategic location at the top of an increasingly organized conservative network’ (Hacker and Pierson 2005b: 135), a network, which is largely organized by wealthy donors who channel considerable fund activists and, in particular, conservative think tanks (Frank 2004: 82).

Theda Skocpol's research on "The shifting U.S. Political Terrain" provides even more detailed insights in the nexus between political and economic interest. Drawing on data on the historical development, funding structure, (political) goals, staff, and network ties, the project revealed how the very same funding organizations of the "Koch network" managed to bypass more traditional channels for political donations, such as party committees, this increasing the financial and political leeway of non-party organizations such as think tanks (Skocpol and Hertel-Fernandez 2016: 683). Most importantly, as Skocpol and Hertel-Fernandez are pointing out, the aspect of value homophily is not only a structural feature of the conservative network but also a transformative force:

> Skocpol and Hertel-Fernandez demonstrate that funding agencies controlled by the Koch brothers were highly successful in orchestrating funding activities with the aim to "reorient" the Republican party by pulling it "toward the ultra-free-market right". (ibid.: 687)

The apparent success of attempts to reorient Republican Party politics can (at least) partly be explained by the concrete advantages an "anti-government/pro-fee-market" rhetoric bring to conservative lawmakers: According to Amy Fried and Douglas Harris, promoting distrust in government allows conservative organizations to develop a coherent narrative and candidates of the political right to run on an anti-elite ticket (Fried and Harris 2001: 159–160). Most importantly Fried and Harris point out that conservative networks of policymakers and purpose-built organizations (like think tanks) can promote "public anger" to moving forward their agenda (ibid.: 167–168).

Likewise, attempts to influence the public opinion can play with a variety of emotions. For instance, in 2006, the Competitive Enterprise Institute launched a large media campaign titled "We call it life" (Boykoff and Rajan 2007: 210). Opposing 'some politicians [who] want to label carbon dioxide a pollutant', the short film explains that CO_2 is a natural gas ('we breathe it out, plants breathe it in') before introducing an alternative claim: 'Carbon dioxide, they call it pollution we call it life'.

No wonder that Greenpeace called the advert "bizarre" (Clarke 2016). However, the failure to produce a "rational" argumentation and play with emotions instead is not a shortcoming but actually a key feature of the success strategy of conservative think tanks: Think tanks like the CEI are instrumental in communicating politics from the top to the lower levels of local activists, movement, and citizens. The message of the campaign connects to anti-government ("some politicians…") and anti-intellectual sentiment by ridiculing the idea that a natural gas that we're breathing out could cause serious dangerous climate change.

From a scientific perspective, the "We call it life" camping seems to be "bizarre" indeed. However, considering that the target audience of the campaign wasn't the scientific community and keeping in mind that conservative think tanks frequently reach out for 'significant segments of the general public' (Dunlap and Jacques 2013: 713), the campaign strategy appears to be carefully developed.

Conservative think tanks in the United States do not primarily engage with climate scientist. Their dissenting positions on climate science systematically evade peer-reviewed processes (Dunlap and Jacques 2013: 701–702) but are directly fed in the news cycle. Conservative think tanks in the United States therefore provide a valuable service to business interest and conservative policymakers. In 2012 an empirical study found that the 'most important factor in influencing public opinion on climate change, however, is the elite partisan battle over the issue' (Brulle et al. 2012: 185). Under these circumstances the role think tanks play is even more important. Think tanks serve as "all-purpose weapons" of and within conservative networks such as the CHC, simultaneously able to challenge climate science (e.g. by setting up a Nongovernmental International Panel on Climate Change), provide expertise to lawmakers (e.g. by holding briefings in Congress), and reach out to the wider public by launching large media campaigns.

Moreover, conservative think tanks do not simply pursue an anti-environmentalist agenda. Challenging climate science is rather another front in an ideological campaign against government regulation. The claims of "mainstream" climate science are contested not because of some scientific dissent but because of their likely political consequences.

In sum, the visible think tanks in the US belong to a conservative network that is structured around a shared belief system that favours free-market politics and a fundamental opposition to government regulation. In the increasingly polarized political landscape of the United States, climate change poses a major threat. However, while believers in climate change would argue that climate change threatens mainly the wellbeing and the survival of future generations, for conservatives, climate change is deemed to be an immediate menace to their way of life.

The next chapter deals with the roles and strategies in a political environment that is in many respects fundamentally different: the Federal

Republic of Germany. Comparing American think tanks to their German counterparts will then allow to identify peculiarities of the two countries and hence arrive at a better understanding of how think tank behaviour is shaped by their respective environment.

Notes

1. Being a prominent climate sceptic in the USA, Fred Singer is particularly active in creating controversies on climate change and climate politics. Tracking his activities Naomi Oreskes and Erik Conway found that Singer's involvement in debates in environmental politics dates back until the late 1970s usually putting him up an environmental regulation (2010: 85–86, 205–207).
2. The report can be obtained free of charge on the NIPCC homepage http://climatechangereconsidered.org/
3. The climategate scandal has considerable influence on the international community of climate scientists itself. In 2014, for instance, climate scientists warned that 'Climate science has become "too green" and "blind to bias"' (Knapton 2014) criticizing the *Environmental Research Letters* for refusing to publish a critical article and warning that this has 'the potential to do as much damage to climate science as the "climategate" scandal, where the University of East Anglia was accused of manipulating data and attempting to suppress critics' (ibid.).
4. Aaron McCright and Riley Dunlap found that '[c]onservative think tanks and their allied climate change contrarians successfully exploited American news media norms—especially the "balancing norm", or the equation of "objectivity" with presenting "both sides of the story"—to achieve a level of media visibility incommensurate with the limited scientific credibility of their claims (McCright and Dunlap 2003; Boykoff and Boykoff 2004). The effectiveness of this strategy is reflected by comparative studies showing that U.S. newspapers are more likely to portray climate science as "uncertain" than are those in other developed nations' (McCright and Dunlap 2011: 159).
5. In February 2015, Republican Senator James "Jim" Inhofe (Oklahoma), author of the 2012 *The Greatest Hoax: How the Global Warming Conspiracy Threatens Your Future*, pulled a snowball from a plastic bag, tossing it to the senate floor in order to disprove global warming as a hoax adding: 'In case we have forgotten, because we keep hearing that 2014 has been the warmest year on record, I ask the chair: You know what this is? It's a snowball, that's just from outside here. So it's very, very cold out. Very unseasonable' (D'Angelo 2017). However Sen. Inhofe's "political stunt" has been widely criticized for its deliberate misunderstanding of the concept of global warming and got considerable media attention, allowing the senator to make his common sense case against global climate change.

6. This depiction is consistent with the findings of Michael Kimmel and Abby Ferber. In their study "White Men Are This Nation" (2000), Kimmel and Ferber explore the normative convictions of (conservative to far-right) militias in rural America. They found that '[c]entral to militia ideology is its antistatist position. Big government, not big capital, is eroding Americans' constitutional right' (Kimmel and Ferber 2000: 593).
7. All calculations are based on the collection of 990 forms provided by conservativetransparency.org/.
8. The GCC founded in 1989 was mainly driven by US-based and large international corporations including economic "heavyweights" such as Ford, BP, and Dupont. While the GCC was actively lobbying for business interest in the run-up to the 1997 Kyoto summit, the front began to crumble already when main contributors (e.g. BP) defected, realizing that adapting to climate regulation might serve business interests better than simply denying human-made climate change (cf. Brown 2000).

References

Abelson, Donald E. 2004. The Business of Ideas: The Think Tank Industry in the USA. In *Think Tank Traditions. Policy Research and the Politics of Ideas*, ed. Diane Stone and Andrew Denham, 215–231. Manchester: Manchester University Press.

———. 2009. *Do Think Tanks Matter? Assessing the Impact of Public Policy Institutes*. 2nd ed. Montreal/Kingston/London/Ithaca: McGill-Queen's University Press.

Antonio, Robert J., and Robert J. Brulle. 2011. The Unbearable Lightness of Politics: Climate Change Denial and Political Politicization. *The Sociological Quarterly* 52 (2011): 195–202.

Bast, Joseph. 2017. Winning the Global Warming War. *The Heartland Institute*, February 17. https://www.heartland.org/news-opinion/news/winning-the-global-warming-war. Accessed 10 Sept 2017.

Baum, Matthew A., and Tim Groeling. 2008. New Media and the Polarization of American Political Discourse. *Political Communication* 25: 345–365.

Borgatti, Stephen P., Ajay Mehra, Daniel J. Brass, and Giuseppe Labianca. 2009. Network Analysis in the Social Sciences. *Science* 323: 892–895.

Boykoff, Maxwell T., and Jules M. Boykoff. 2004. Balance as Bias: Global Warming and the US Prestige Press. *Global Environmental Change* 14: 125–136.

Boykoff, Maxwell T., and Shawn K. Olson. 2013. Wise Contrarians: A Keystone Species in Contemporary Climate Science, Politics and Policy. *Celebrity Studies* 4 (3): 276–291.

Boykoff, Maxwell T., and Ravi S. Rajan. 2007. Signals and Noise. Mass-Media Coverage of Climate Change in the USA and the UK. *EMBO Reports* 8 (3): 207–211.

Brown, Lester R. 2000. The Rise and Fall of the Global Climate Coalition. *Earth Policy Institute*, July 25. http://www.earth-policy.org/plan_b_updates/2000/alert6

Brown, Susan Stamper. 2015. Whatever Happened to Common Sense Regarding Climate Change? *The Christian Post*, June 23. http://www.christianpost.com/news/whatever-happened-to-common-sense-regarding-climate-change-140744/. Accessed 9 Sept 2017.

Brulle, Robert J. 2014. Institutionalizing Delay: Foundation Funding and the Creation of U.S. Climate Change Counter-Movement Organizations. *Climate Change* 122 (4): 681–694.

Brulle, Robert J., Jason Carmichael, and Craig J. Jenkins. 2012. Shifting Public Opinion on Climate Change: An Empirical Assessment of Factors Influencing Concern Over Climate Change in the U.S., 2002–2010. *Climate Change* 114 (2): 169–188.

Butos, William N., and Thomas J. McQuade. 2015. Causes and Consequences of the Climate Science Boom. *The Independent Review* 20 (2): 165–196.

Clarke, Joe Sandler. 2016. Watch the Bizarre Advert Praising CO_2 That Donald Trump's New Climate Guy Helped Produce. November 18. https://unearthed.greenpeace.org/2016/11/18/donald-trump-myron-ebell-cei-climate-change/. Accessed 20 Oct 2017.

D'Angelo, Chris. 2017. 2 Years Ago, Sen Jim Inhofe Pulled This Embarrassing Stunt. *The Huffington Post*, February 26. http://www.huffingtonpost.com/entry/two-year-anniversary-jim-inhofe-snowball_us_58b07bb1e4b060480e07959d. Accessed 9 Sept 2017.

Dunlap, Riley E., and Peter J. Jacques. 2013. Climate Change Denial Books and Conservative Think Tanks. Exploring the Connection. *American Behavioral Scientist* 57 (6): 699–671.

Dunlap, Riley E., and Aaron M. McCright. 2010. Climate Change Denial: Sources, Actors and Strategies. In *Routledge Handbook of Climate Change and Society*, ed. Constance Lever-Tracy, 240–259. Abingdon: Routledge.

———. 2011. Organized Climate Change Denial. In *The Oxford Handbook of Climate Change and Society*, ed. John S. Dryzek, Richard B. Norgaard, and David Schlosberg, 144–160. Oxford: Oxford University Press.

Feldman, Lauren, Edward M. Maibach, C. Roser-Renouf, and Anthony Leiserowitz. 2012. Climate on Cable: The Nature and Impact of Global Warming Coverage on Fox News, CNN and MSNBC. *The International Journal of Press/Politics* 17 (1): 3–31.

Fisher, Gene H., and Warren E. Walker. 1994. Operations Research and the RAND Corporation. *RAND Corporation* Paper No. 7857. https://www.rand.org/pubs/papers/P7857.html

Fortun, M., and S. Schweber. 1993. Scientists and the Legacy of World War II: The Case of Operations Research (OR). *Social Studies of Science* 23: 595–642.

Frank, Thomas. 2004. *What's the Matter with Kansas? How Conservatives Won the Heart of America*. New York: Metropolitan Books.

Fried, Amy, and B. Harris Douglas. 2001. On Red Capes and Charging Bulls: How and Why Conservative Politicians and Interest Groups Promoted Public Anger. In *What Is It About Government That Americans Dislike?* ed. John R. Hbbing and Elizabeth Theiss-Morse. Cambridge: Cambridge University Press.

Government Response to the House of Commons Science and Technology Committee Report. 2010. The Disclosure of Climate Data from the Climatic Research Unit at the University of East Anglia. Cm 7934.

Hacker, Jacob S., and Paul Pierson. 2005a. Abandoning the Middle: The Bush Tax Cuts and the Limits of Democratic Control. *Perspectives on Politics* 3 (1): 33–53.

———. 2005b. *Off Center. The Republican Revolution and the Erosion of American Democracy*. New Haven/London: Yale University Press.

Hein, James Everett, and Craig J. Jenkins. 2017. Why Does the United States Lack a Global Warming Policy? The Corporate Inner Circle Versus Public Interest Sector Elites. *Environmental Politics* 26 (1): 97–117.

Hofstadter, Richard. 1963. *Anti-intellectualism in American Life*. New York: Alfred A. Knopf.

Iyengar, Shanto, and Kyu S. Hahn. 2009. Red Media, Blue Media: Evidence of Ideological Selectivity in Media Use. *Journal of Communication* 59: 19–39.

Kimmel, Michael, and Abby Ferber. 2000. White Men Are This Nation: Right-Wing Militias and the Restoration of Rural American Masculinity. *Rural Sociology* 65 (4): 582–604.

Klein, Naomi. 2011. Capitalism vs. the Climate. *The Nation*, November 9. https://www.thenation.com/article/capitalism-vs-climate/. Accessed 8 Sept 2017.

———. 2014. *This Changes Everything. Capitalism vs. the Climate*. New York/London/Toronto/Sydney/New Delhi: Simon & Schuster.

Knapton, Sarah. 2014. Climate Change Science Has Become 'Blind' to Green Bias. *The Telegraph*, May 16. http://www.telegraph.co.uk/news/science/science-news/10837146/Climate-change-science-has-become-blind-to-green-bias.html. Accessed 2 Sept 2017.

Kuo, Didi, and Nolan McCarthy. 2015. Democracy in America, 2015. *Global Policy* 6 (S1): 49–55.

Lakoff, George. 2002. *Moral Politics. How Liberals and Conservatives Think*. Chicago/London: The University of Chicago Press.

Mayer, J. 2016. *Dark Money: The Hidden History of the Billionaires Behind the Rise of the Radical Right*. New York: Doubleday.

McCright, Aaron M., and Riley E. Dunlap. 2003. Defeating Kyoto: The Conservative's Movement Impact on U.S. Climate Change Policy. *Social Problems* 50 (3): 348–373.

———. 2011a. The Politicization of Climate Change and Polarization in the American Public's View of Global Warming 2001–2010. *The Sociological Quarterly* 52: 155–194.

———. 2011b. Cool Dudes: The Denial of Climate Change Among Conservative White Males in the Unites States. *Global Environmental Change* 21: 1163–1172.

McDonald, Lauren. 2014. Think Tanks and the Media. How the Conservative Movement Gained Entry into the Education Policy Arena. *Educational Policy* 28 (6): 845–880.

McKitrick, Ross. 2001. What's Wrong with Regulating Carbon Dioxide Emissions? Briefing at the United States Congress, October 11.

McPherson, Miller, Lynn Smith-Lovin, and James M. Cook. 2001. Birds of a Feather: Homophily in Social Networks. *Annual Review of Sociology* 27: 415–444.

Merton, R. 1973. The Normative Structure of Science. In *The Sociology of Science. Theoretical and Empirical Investigations*, ed. Robert K. Merton, 267–278. Chicago/London: University of Chicago Press.

Mirowski, Philip. 2002. *Machine Dreams. Economics becomes a Cyborg Science*. Cambridge: Cambridge University Press.

Mirowski, Philip, and Dieter Plehwe. 2009. *The Road from Mont Pélerin: The Making of a Neoliberal Thought Collective*. Cambridge, MA: Harvard University Press.

Oreskes, Naomi, and Erik Conway. 2008. Challenging Knowledge: How Climate Science Became a Victim of the Cold War. In *Agnotology. The Making and Unmaking of Ignorance*, ed. Robert N. Proctor and Londa Schiebinger, 55–89. Stanford: Stanford University Press.

———. 2010. *Merchants of Doubt. How a Handful of Scientists Obscured the Truth on Issues from Tobacco Smoke to Global Warming*. London/New Delhi/New York/Sydney: Bloomsbury.

Powell, James Lawrence. 2011. *The Inquisition of Climate Science*. New York: Columbia University Press.

Republican Party Platform. 1984. America's Future Free and Secure. http://www.presidency.ucsb.edu/ws/index.php?pid=25845. Accessed 15 Oct 2017.

Rich, Andrew. 2001. U.S. Think Tanks and the Intersection of Ideology, Advocacy, and Influence. *NIRA Review: A Journal of Opinion on Public Policy Worldwide* 8: 54–59.

———. 2004. *Think Tanks, Public Policy and the Politics of Expertise*. Cambridge/New York: Cambridge University Press.

Rousseas, Stephen. 2016. *The Political Economy of Reaganomics: A Critique*. London/New York: Routledge.

Ruser, Alexander, and Amanda Machin. 2017. *Against Political Compromise. Sustaining Democratic Debate*. London: Routledge.

Skocpol, Theda, and Alexander Hertel-Fernandez. 2016. The Koch Network and Republican Party Extremism. *Perspectives on Politics* 14 (3): 681–699.

Trefethen, F. 1954. A History of Operations Research. In *Operations Research for Management*, ed. Joseph F. McCloskey and Florence N. Trefethen, 3–35. Baltimore: Johns Hopkins University Press.

Weaver, K. 1989. The Changing World of Think Tanks. *Political Science and Politics* 22 (3): 563–578.

Weber, Elke U., and Paul C. Stern. 2011. Public Understanding of Climate Change in the United States. *American Psychologist* 66 (4): 315–328.

Wells, H.G. 1914. *The World Set Free. A Story of Mankind*. New York: E.P. Dutton & Company Publishers.

Williamson, Vanessa, Theda Skocpol, and John Coggin. 2011. The Tea Party and the Remaking of Republican Conservatism. *Perspectives on Politics* 9 (1): 25–43.

CHAPTER 5

Members Only: Think Tanks and Climate Politics in Germany

Introduction

Otto von Bismarck, chancellor of the German Empire and architect of modern Germany, didn't expect too much from scientific experts and advisors: 'There is no exact science of politics just as there is none for political economy. Only professors are able to package the sum of the changing needs of cultural man into scientific laws' (Otto von Bismarck cited after Pflanze 1968: 88).

Indeed 'Professor was a very nasty word in Bismarck's vocabulary' writes Otto Pflanze (1968: 88) explaining the Prussian statesman's aloofness towards expert advice. 'Politics is less a science than an art' thought Bismarck adding that '[i]t is not a subject which can be taught. One must have the talent for it. Even the best advice is of no avail if improperly carried out' (ibid.: 89).

What role could think tanks play in a political environment that considers politics an art and values "talent" rather than knowledge? To understand the evolution of German think tanks, one has to take their starting conditions into account. Moreover, although think tanks are now numerous in the Federal Republic of Germany (a count based on the "think tank directory"—and the author's own research identifies 153 active think tanks in 2017), assessing their role and impact isn't possible without considering the political landscape, the intellectual climate, and an "etatist approach" (Esping-Andersen 1990: 40) of direct state-intervention to grant social rights and solve social and economic problems.

A. Ruser, *Climate Politics and the Impact of Think Tanks*,
https://doi.org/10.1007/978-3-319-75750-6_5

This chapter will start with a brief outline of the history of think tanks in Germany. Tracing the "evolution" of think tanks in Germany is crucial for fully understanding how they fit in the political and academic environment and, ultimately, why German think tanks differ from their American counterparts.

At first glance, pointing to the differences seems to be relatively simple: Compared to the USA, the number of think tanks is small. Moreover, unlike American think tanks which are predominantly based in Washington D.C., German think tanks are spread all over the Federal Republic. ALos, Berlin (or, previously Bonn) doesn't resemble the "battleground of ideas" that is the capital of the United States. Another more obvious difference is the issue orientation of think tanks: 'Most think tanks in Germany are neither single-issue institutes nor full-service institutions' writes Martin Thunert (2004: 80), thus indicating that cooperation and competition between German think tanks follow rules distinctly different from the US.

Even a cursory examination of think tanks in Germany reveals even more differences to the US. As will be discussed in more detail below, state-sponsored networks of independent research institutes, the *Helmholtz Association of German Research Centres* and the *Leibniz Association*, provide the most important frameworks, providing think tanks with financial resources and access to public authorities.

At the same we would find that the dominance of publicly funded academic think tanks doesn't imply that partisanship and advocacy are all but absent. A special type of think tank "political" (or "partisan") foundation plays an important role, and it is no coincidence that the Friedrich-Ebert-Foundation, closely allied with the Social Democratic Party, was one of the first policy-oriented research organizations to be established in Germany.

This being said it might seem odd that scholars seem to agree that 'Germany is not the best-researched country when it comes to think tanks' (Pautz 2010: 278). But why?

Is it because think tanks have generally been overlooked by the scientific community? Or are German think tank researchers lacking behind since the spread of think tanks in the Federal Republic of Germany is a rather recent phenomenon? Or is it for the simple reason that think tanks are less important, are less influential, and, in consequence, are less interesting?

The apparent neglect of German think tanks seems the more surprising since there are no recent phenomenon: 'In the last two decades of the twentieth century, think tanks proliferated dramatically. Countries where think tanks were already present such as the USA, Britain, Sweden, Canada, Japan, Austria and Germany witnessed further organisational growth' (Stone 2007: XX).

To understand why (advocacy) think tanks haven't attracted much scholarly attention, one has to consider the historical starting conditions and trajectories of public policy research in Germany. For there wasn't a lack of interest in policy research, policy researchers, or the role of advice. Scholarly attention had rather been channelled away from private research institutes. Instead state-sponsored, public research institutes with close ties to public universities and public authorities and most notably political foundations had been at the centre of attention (Pogorelskaja 2002; Speth 2010; von Arnim 1991; von Vieregge 1977, 1990).

Listing these differences is easy. However, understanding why German think tank look differently and, perhaps, act in a different way requires to taking into account different political traditions and a different intellectual climate.

Born Out of Pessimism? Early Think Tank in Germany

What was the "intellectual climate" of the time the first German think tanks were created? Apart from strong etatist traditions which favoured state intervention for solving social problems, Germany at the turn of the twentieth century seemed to be characterized by what was called "cultural pessimism". Etatist tradition explains the driving forces behind the foundation of the first think tanks that didn't come from the private sector. In contrast to the United States, it wasn't the effort of private donors but the Prussian State and the public authorities of the Weimar Republic which set up and fund these first academic think tanks. Close ties between publicly funded think tanks and the state dominate the German think tank landscape to the present day.

German think tank dates back to the German Empire. The first think tanks, the *Hamburgisches Welt-Wirtschafts Archive* (*HWWA*) (Hamburg Archive of Global Archive) and the *Institut für Weltwirtschaft* (Institute for the World Economy), were founded in 1908 and 1914, respectively, to provide solutions to global economic problems. Think tanks in Germany first began to prosper in the Weimar Republic. Between 1925 and 1927, five think tanks, the *Friedrich-Ebert-Foundation* (1925), the *Deutsche Institut für Wirtschafttsforschung* (*DIW*) (1925), the *Rheinisch-Westfälische*

Institut für Wirtschaftsforschung (1926), the *Arbeitsgemeinschaft für wirtschaftliche Verwaltung* (1926), and the *Finanzwissenschaftliche Forschungsinstitut Cologne* (1927), were founded to conduct economic and social policy research.

While the Friedrich-Ebert-Foundation maintained close ties to the Social Democratic Party, the other four research institutes founded in 1926 and 1927 had close ties to government authorities and the academic system of the Weimar Republic.

The DIW, for instance, was founded by Ernst Wagemann, then director of the Central Statistical Office (*Statistisches Reichsamt*), and the Rheinisch-Westfälische Institut für Wirtschaftsforschung and the Finanzwissenschaftliche Forschungsinstitut were directly affiliated to the academia. The HWWA, the oldest of the institution, was created as the central department of the *Hamburgisches Kolonialinstitut* (Hamburg Colonial Institute) which developed into the Hamburg University after World War I.

The founding of these research facilities as *public* policy research institutes was consistent with attempts to make political and economic use of state-of-the-art research: 'Weimar Germany was marked by ambitious attempts to extend science and technology into everyday life under conditions of political crisis' (Hopwood 1996: 117).

In addition to the etatist tradition which explains the organizational structure of early German think tanks, a pessimistic intellectual climate, especially after World War I, shaped their agenda: In contrast to the United States, the intellectual climate in the Weimar Republic was rather hostile to science: 'After the Great War, German scientists lost much of their prestige; Spengler had just published his widely popular *Decline of the West* and Spenglerism was everywhere' writes James R. Brown (2001: 116). Unlike the United States accelerated technological progress, scientific breakthroughs, and social change weren't heralding a Progressive Era. Spenglerism (among other things) was both a symptom of and a catalyst for a "cultural pessimism" (Kaes et al. 1994: 355) and scepticism towards science (Frye 1974: 13). Rather than contributing to the "improvement" of society and decision-making, German scientist were expected to contribute to slowing down or halt the decline of culture. Far from promoting social progress (social), scientists were expected to preserve cultural and social achievements.

Moreover, Germany wasn't only lacking the intellectual climate but also the sponsors for the development of large-scale, independent research organizations. The civil society in the Weimar Republic wasn't dominated by philanthropic entrepreneurs, wasn't driven by private donation and charity. Provision of public goods, social welfare, and solutions for social problems obliged the state. Thomas Adam notes that 'the German bourgeoisie did not develop feelings of responsibility for German society; rather (…) they expected the state to take responsibility for financing social and cultural institutions' (Adam 2004: 3).

Think Tanks in the Federal Republic of Germany

The rebuild of the West German political system after the Second World War was driven by an optimistic, German version of the Progressive Era. The era of the so-called Planungseurophorie ("planning euphoria") in the 1950s and 1960s (Thunert 2001: 223) was characterized by the conviction that political decision-making in Germany is increasingly dependent on expert advice (Rufloff 2004: 179). Academic advisory councils were set up as early as 1948 that is well before the founding of the Federal Republic of Germany (Rufloff 2004: 178). Composed exclusively of renowned scientists, the main task of these councils was to provide general guidance and orientation for the public administration and the ministries in the newly founded political system (ibid.).

Policy advice was and is seen as a service provided to regional and federal ministries and parliaments. In-house capacities such as the *Wissenschaftlichen Dienste des Bundestages* (Research Service of the German Bundestag) do politically "neutral" research by request of the political parties in the federal parliament, thus informing parliamentary debate and legislation (for a detailed description, see von Winter 2006). In addition, "sections" (*Refereate*) within ministries employ their own experts to build in-house capacities for policy research and analysis.

Moreover, the federal government was instrumental in setting up independent research capacities such as the *Stiftung Wissenschaft und Politik* (*SWP*) (Science and Politics Foundation) which was deliberately emulating the American RAND Corporation (Thunert 2004: 73).

Apart from these rather rare initiatives by public authorities, there was no rapid growth of think tanks in Germany. Unlike in the United States the 1970s and 1980s saw a steady but slow increase in their number. As displayed in Fig. 5.1, the number of think tanks increased continuously, peaking in the first half of the 1990s and then again in the early 2000s.

As can be seen in Fig. 5.1, the years between the founding of the Federal Republic in 1949 and the German Reunification in 1989/1990 witnessed only a slow growth in the number of think tanks. Even more important than their relative small number was their limited influence on policymaking in West Germany. Summarizing the impact of West German think tanks, Renate Mayntz found in 1987 that '[t]here are some (…) policy research institutes which serve the government collectively, but the influence of these bodies on government policy is mostly rather indirect and it would be difficult to trace specific policy decisions to their advice' for 'the West Germany system has relatively little by way of a specialized infrastructure for policy analysis and advice' (Mayntz 1987: 8–9).

Since the German Reunification, the number of think tanks has increased (Jochem 2013: 231). However, so far no fundamental changes in the composition, the roles, and the strategies of German think tanks can be observed. Publicly funded, relatively large academic think tanks are still dominating the think tank landscape. Partisan expertise is mainly provided by political foundations. Since the founding of the Federal Republic, all established political parties in Germany followed the example of the Social Democratic Party (SPD) and set up political foundations of their own. The *Konrad-Adenauer-Foundation*, close to the Christian Democratic

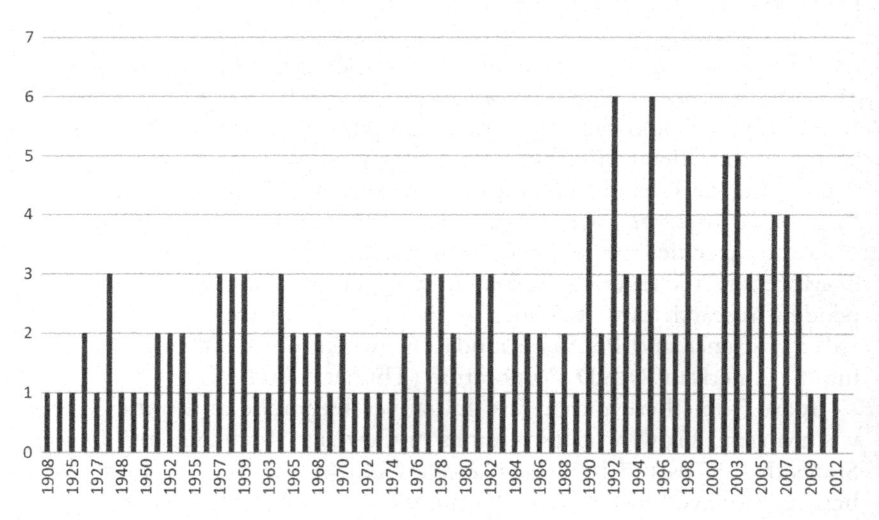

Fig. 5.1 Foundation of think tanks in Germany (1949–1989 West Germany only). (Source: Think Tank Directory, authors' calculations)

Union (CDU), was founded in 1955. The liberal *Friedrich-Naumann-Foundation* tied to the Liberal Party (FDP) followed in 1958. The conservative party in Bavaria (CSU) traditionally in coalition with the CDU founded its own political foundation, the *Hans-Seidel-Foundation*, in 1967. The *Rosa-Luxemburg-Foundation* came into being shortly after the German Reunification to provide expertise and support to the newly founded Left Party (then "Partei des Demokratischen Sozialismus, PDS" later renamed to "Die Linke"). Of the established parties the Green Party was the last to establish a political foundation. It wasn't before 1997 that the *Heinrich-Böll-Foundation* was created. Most recently right-wing "Alternative for Germany" (AfD) is attempting to start a political foundation by its own. Although overshadowed by internal conflict and legal concerns (cf. Burchard 2015), setting up a political foundation close to the AfD could, for the first time, bring in an advocacy organization which is actively denying climate change (see below).

The activities of political foundations are one important aspect limiting the influence of advocacy think tanks. The lasting ties between political foundations and political parties systematically decrease the demand for the services of advocacy think tanks. Another aspect which makes the German political system particularly difficult for privately funded, non-academic think tanks is the existence of institutionalized networks/frameworks of academic think tanks and research institutes.

In 1991 representatives of 32 extramural research institutes formed the "Arbeitsgemeinschaft Blaue Liste" (*Blue List*). Renamed in 1997 the "Leibniz Association" in 2016 has 88 member institutes which employ a total of 18.668 people (9.485 researchers). In the same year member institutes received 384.16 million euro in external grants (21 per cent of the total budget) and a total of 1.076 billion euro in public funds (cf. https://www.leibniz-gemeinschaft.de/ueber-uns/leibniz-in-zahlen/). Member institutes do not only have access to substantial public funding but are granted privileged access to policymakers. Member institutes are commissioned to issuing economic report to the government twice annually thus heavily influencing economic and fiscal politics in Germany (Thunert 2004: 74).

Founded in 1995 the *Helmholtz-Gemeinschaft Deutscher Forschungszentren* (Helmholtz Association) consists of 18 research institutes. The mission of the Helmholtz Association is 'to solving grand challenges facing society, science and industry by conducting top-rate research in the fields of Aeronautics, Space and Transport; Earth and Environment;

Energy; Health; Matter; and Key Technologies' (https://www.helmholtz.de/en/about_us/the_association/).

In 2017 member institutes command a budget of 4.38 billion euro. About two thirds are public funds with additional money stemming from grants, cooperation, and licence fees. A total of 38.733 employees work for member institutes. The Helmholtz Association maintains ties and cooperates closely with German universities (cf. https://www.helmholtz.de/ueber_uns/die_gemeinschaft/zahlen_und_fakten/).

Moreover, its governance structure reveals close and lasting ties between the Association and public authorities: Its managing body (the "Senate") consists of the federal minister for research and education, two ministers for research from the regional states (*Länder*), six scientists, six representatives of economy, one representative of the federal ministry of finance, one representative of a regional ministry of finance, two representative of another research association (e.g. the German Research Association, but also the Leibniz Association), two members of parliament (*Bundestag*), and the president of the Helmholtz Association.

The Helmholtz and Leibniz Associations provide an institutionalized, exclusive framework for extramural research and policy advice. In conjunction with two academic bodies, the Max-Planck-Society,[1] the Fraunhofer Society,[2] and the Helmholtz and Leibniz Associations are setting the tone for policy-relevant research in the Federal Republic of Germany.

In 2005 the federal and regional governments adopted the *Joint Initiative for Innovation and Research* ("Pakt für Forschung und Innovation") which aims at guiding the future direction of research, provides a safe and reliable environment for extramural research, and fosters the convergence of research objectives and research policy goals.

The Joint Initiative is supported by and tailored to the needs of the Leibniz Association, the Helmholtz Association, the Max-Planck-Society, Fraunhofer Society, and the German Research Association. Annual monitoring reports and exchange guarantee a close dialogue between the researchers and policymakers. But the "Joint Initiative" does not only provide a stable environment. Participating in it provides member institutes of the four associations and societies with additional influence and leverage. For instance, as documented in the 2015, monitoring report member institutes (e.g. the "Institut für Weltwirtschaft", member of the Leibniz Association) are coordinating research projects and policy dialogues on climate change (GWK 2015: 467).

By deciding on research foci and by maintaining close, institutionalized ties to government bodies' research associations and societies participating in the Joint Initiative have a clear advantage. Their privileged access to policymakers at the federal and regional level effectively raises the bar for new, alternative providers of political expertise. Since member organizations adhere to the standards of good scientific practice getting access is even more difficult for non-academic advocacy think tanks.

It is important to keep this pre-structuring of the research landscape in mind to understand the more recent development of think tanks in Germany.

Of the 153 think tanks registered in the Federal Republic of Germany (see thinktankdirectory database[3]), the majority can be labelled as academic think tank with advocacy think tanks account for only one quarter (37 organizations, i.e. 24.34 per cent).[4] Almost a third (11 organizations, that is, 29.72 per cent) of these advocacy think tanks have been created after the year 2000 indicating a considerable shift in the mix of German think tanks in recent years. This still leaves the question of whether the numerical increase is accompanied by an increasing importance of advocacy think tank.

In the following paragraph the role and impact on national climate politics will be examined more thoroughly.

Climate Think Tanks and National Climate Politics in Germany

Germany likes to portray itself as a frontrunner in international climate protection as well as a model for national climate politics (Hustedt 2013: 96). People in Germany are on average more concerned about climate change (according to a representative survey conducted by PEW research in 2015, 55 per cent in Germany but only 45 per cent of people in the USA believe global climate change to be a "very serious problem", PEW 2015: 13),[5] and until very recently climate sceptic parties have been all but absent on the political stage.

The federal structure of Germany, the need to form party coalitions at the regional and the federal level, and the strong position of the ministerial bureaucracies in the decision-making process favour consensus orientation and political compromise (see Chap. 3).

Climate politics directly relates to one of the "objectives of the state". Article 20a of the basic law of Germany (*Protection of the natural foundations of life and animals*) reads:

> 'Mindful also of its responsibility toward future generations, the state shall protect the natural foundations of life and animals by legislation and, in accordance with law and justice, by executive and judicial action, all within the framework of the constitutional order.' Since lawmakers at the regional and federal level are legally bound to protect the livelihood of future generations, the necessity of climate politics isn't controversial in the Federal Republic.

The Impact of the 2006 Act of the Amendment of Basic Law on National Climate Legislation

The complexity of the lawmaking process, in particular, the division of power between the regional and the federal levels while guaranteeing a maximum of exchange between various jurisdictions, was nevertheless seen as hampering the federal government's ability to 'enact comprehensive and uniform regulations' (http://www.umweltbundesamt.de/en/the-german-environmental-constitutional-law). Consequentially, redistribution of legislative power in environmental politics was an important component of the 'Act for the Amendment of the Basic Law' (Federal Law Gazette I 41: 2034):

> Of particular interest for environmental law was the redistribution of legislative powers for the protection of the environment. Prior to the reform, the Basic Law shared the legislative powers of the Federal Government in respect of environmental protection across various, mostly non-environment-specific jurisdictions. They came under either the concurrent or the framework legislative competence of the Federal Government. (http://www.umweltbundesamt.de/en/the-german-environmental-constitutional-law)

The changes in the legal structure and the lawmaking process reflect a growing political consensus on climate science and climate politics since the 1980s. The increasing awareness of environmental problems created a political demand for scientific expertise. Universities and public research institutes were quick to cater to these demands:

> Germany, it seemed, converted almost overnight "from laggard to leader" (…). By the 1990s, the German climate research system had become one of

> the best-equipped in the world, and the country had established a reputation as one of the political pacemakers in the international arena, particularly for its ambitious goals for the reduction of CO2 emissions. (Krück et al. 1999)

The "best-equipped" research system is structured around research frameworks such as the Helmholtz Associations "Atmosphere and Climate Programme". The goal of this programme 'is to better understand the function of the atmosphere within the climate system. For this purpose scientists will carry out extensive measurements of atmospheric parameters, perform laboratory tests and create numerical models of processes that play an important role in the atmosphere' (https://www.helmholtz.de/en/research/earth_and_environment/atmosphere_and_climate/).

The research programmes of the Helmholtz and Leibniz Associations commensurate with the state of research of the international community of climate scientist. In fact, Ottmar Edenhofer, deputy director and chief economist of the Potsdam Institute for Climate Impact Research, served as co-chair of an IPPC Working Group from 2008 to 2015.

Networks of Influence: Status, Social Capital, and the Impact on National Climate Politics

The previous paragraphs dealt with some key aspects of the institutional environment academic think tanks operate in. It was shown that climate think tanks such as the Potsdam Institute are important members of the international scientific community and contribute to "mainstream" climate science. Moreover, member organizations of the Helmholtz and Leibniz Associations maintain close ties to ministries and the Federal Environment Agency.

It can be assumed that these think tanks are in a privileged position to influence national climate politics. However, in order to estimate the relative importance of academic think tanks which belong to these associations, they have to be compared to other academic and advocacy think tanks.

Analogous to Chap. 4, this relative importance should be estimated by determining their position within a network. Network data was compiled by combining information from the "think tank directory database" (basic information on think tank types) and thinktankmap.org. Using an inductive method, websites of the think tanks listed in the abovementioned databases were searched for past and ongoing research and advisory projects. The data extracted include information on clients and cooperation partners and allowed calculating an undirected network of climate-related research cooperation. The final data file included information on an ongoing project cooperation (year of reference is 2013) as well as organizational data (type of the respective organization) and was analysed by using the SNA software Pajek to uncover ties between contractors (think tanks) and clients. Twenty-one academic and 12 advocacy think tanks were included in the data set. In addition 13 government authorities, occurring as clients, and four industrial/civil society organizations (clients/cooperation partners) were considered. The ratio of academic and non-academic think tanks reproduces structural data provided by Thunert (2004) who describes the German case as dominated by academic think tanks (accounting for about 50 per cent) with advocacy think tanks (about 30 per cent) trailing behind (contract research institutions and party affiliated think tanks accounting for 10–15 per cent each) (Thunert 2004: 72).

The most important result of the network analysis concerns the relative importance of academic think tanks in Germany: In contrast to the US *academic* think tanks are the key players. To measure the relative importance of all actors in the network, the respective closeness centrality degrees were calculated. Closeness centrality is calculated on the basis of the length of the average shortest path between a node (actor) and all other actors in a network graph and indicates how "easy" or quick an actor can reach (e.g. for spreading information) other actors within the network and is therefore a suitable measure for the social capital of an actor.

Focusing on the closeness of the actor revealed significant differences between advocacy and academic think tanks: While the average closeness, centrality is almost equal (0.26 / advocacy 0.27 academic think tanks) a privileged group of academic think tanks could be identified. The most central actors in the network are the *Potsdam Institute for Climate Impact Research* (0.4353), the *Wuppertal Institute for Climate, Environment and Energy* (0.3718), and the *Institute for Ecological Economy Research* IÖW (0.3499). All three are academic think tanks with the *Öko-Institute* (0.2950) as the most important (pro-climate) advocacy think tank trailing behind. As displayed in Fig. 5.2, governmental authorities also occupy a central position. Ministries and federal agencies are by far the most important clients for environmental think tanks. Funding lines and project funding administered or provided by ministries and the Federal Environment Agency are by far the most important sources of research

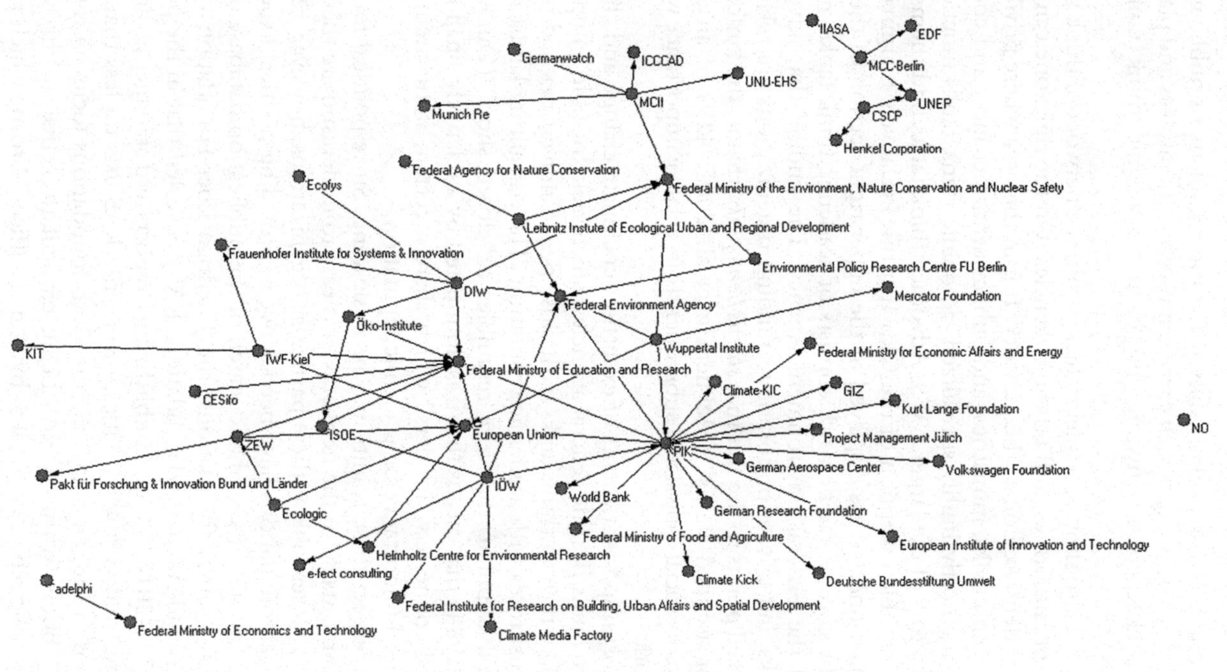

Fig. 5.2 Network climate research, think tanks, clients. (Source: Own calculation)

funding. The network analysis shows that the recipients of public grants are academic think tanks, which carry out projects themselves and play the role of gatekeepers by inviting less central academic think tanks to cooperate.

Figure 5.2 also shows the hierarchies within the network of think tanks, government authorities, and funding agencies. On the right one can see a particularly dense region of the network. It is in this area where privileged academic think tanks, ministries, and public authorities meet and cooperate. Smaller, mostly highly specialized, academic think tanks function as sub-contractors of the larger UWS. Advocacy think tanks find themselves (with the notable exception of the Öko-Institute) relatively marginalized or even disconnected (as displayed in the upper right section of the network). The core of the network consists of academic think tanks, ministries, and the Federal Environment Agency. The finding that advocacy think tanks and private or corporate funding organizations are relatively marginalized points to the fact that *status homophily*, that is, the "informal, formal, or ascribed status" (McPherson et al. 2001: 419) of an actor, determines its social capital and subsequently its relative importance within the network.

The most important clients of expertise on climate change and climate politics are ministries and federal agencies which seek "reliable", impartial advice. They turn to think tanks with close ties to universities and public research institutes with academic reputation. The consulting landscape in the Federal Republic therefore resembles a "closed shop" (Dunn and Gennard 1984) that means that it is dominated by a relatively small number of well-connected "insiders" which share a certain status stemming from political importance or academic reputation.

In one respect these results may be misleading. Since political foundations remain outside of the networks, it is easy to underestimate their significance. A closer look at these particular organizations, however, reveals that political foundations have been active in providing political expertise on climate politics. A systematic search of the publication database of the two largest political foundations, the Friedrich-Eber-Foundation (FES) and the Konrad-Adenauer-Foundation (KAS), revealed that in the period from 1997 to 2017, the FES published 91 reports and statements directly to "climate" and "climate politics," while the KAS lists no less than 663 publications in the same time period. Both foundations focus on reports and policy briefs (with the KAS being particularly active in publishing country-specific reports compiled by branch offices abroad) which target a national and international audience. Reports and policy recommendations are frequently translated to English, French, Spanish, or Arabic to increase the readership. Moreover, political foundations maintain ties to key members of the network described above. For instance, in 2007, the FES launched the comprehensive "Kompass 2020" project which aimed at providing expertise on the future development of climate politics in Germany and at the global stage. The FES worked closely with Hermann E. Ott, then Head of Berlin Branch Office of the Wuppertal Institute, who also authored the key policy paper on the challenges of international climate politics (Ott 2007). Moreover, in February 2017, the KAS and the Potsdam Institute jointly organized a conference on "Compensating the Costs of Climate Change" in Hong Kong to discuss new ways of mitigating and adapting to anthropogenic climate change (see: http://www.kas.de/wf/de/17.71181/).

Overall, because of the "close personal and ideological links to their mother party" (Thunert 2004: 80), the positions of political foundations are largely consistent with the respective party position, thus reflecting the general consensus on climate change.[6]

Think Tanks in the Media

Moreover, think tanks may not only operate outside of main funding structures but might also address different target audiences. In order to estimate their outreach to the wider public, appearances in print media (including online media) were analysed. Drawing on the "nexis" database (www.nexis.com) which provides information and access to newspapers, periodicals, press releases, and other outlets, climate-related media appearances of think tanks in Germany were identified and counted (see Table 5.1).

The findings confirm the relative importance of the Potsdam Institute for Climate Impact Research and the Wuppertal Institute for Climate, Environment and Energy which rank second and fourth, respectively. However, with regard to media attention, advocacy think tanks, most notably the Öko-Institute, which leads all think tanks in media appearances by a wide margin and Ecofys, are equally visible. However, the advocacy think tanks are *not* challenging the dominant discourse: In a recent study on "Climate Change Communication in Germany", Mike Schäfer found that 'climate change coverage in Germany adopts an "anthropogenic climate change as a global problem frame' (2016: 12).

Table 5.1 Environmental think tanks in the media, 01/01/2005 to 31/12/2016

Think tank	*Newspaper*	*Industry press*	*Periodicals*	*Press release*	*Online*	*Total*[b]
Öko-Institute[a]	2850	806	671	406	301	3951
PIK	2106	267	240	368	327	2959
Ecofys[a]	1275	707	280	676	271	2561
Wuppertal Institute	1353	284	212	107	111	1685
IÖW	198	70	43	28	14	275
Leibniz Institute of Ecological Urban and Regional Development	141	17	13	1	3	158
Helmholtz Centre for Environmental Research	85	4	10	12	17	124
MCC Berlin	48	31	8	54	16	123
Adelphi Research[a]	60	24	8	60	31	115
Ecologic Institute	34	17	8	12	39	87
Climate Media Factory[a]	2	7	1	15	1	19
ISOE	15	3	3	0	1	19
Germanwatch[a]	7	2	2	1	0	8

[a]Advocacy think tanks
[b]Due to multiple usage (e.g. press releases cited in articles), numbers in the table cells exceed the total
Source: Own calculation based on nexis database

The "mission statement" of the Öko-Institute is consistent with this depiction of climate change:

> The aim of our work is to contribute to the preservation of the environment and of natural resources, and to ensure the foundation on which all human life depends, for present and future generations. Oriented at finding solutions, we apply our ideas, our scientific expertise and our consulting capabilities to initiate and form the necessary transformations of policy and society. We believe that such transformation processes must be democratic and equitable – also on the international level. (https://www.oeko.de/en/the-institute/mission-statement/)

Likewise, in its "story" Ecofys describes itself as an organization that "develops innovative solutions and strategies to support its clients in moving forward in the energy transition and the challenge of climate change". (https://www.ecofys.com/en/page/our-story/)

In sum, although advocacy think tanks have access to mass media in Germany, they play a distinctively different role compared to their American counterparts. Since the German media landscape is missing a "dismissive segment" (Schäfer 2016: 18), advocacy think tanks operate in the same consensual setting.

Conclusion: Privileged Access and Consensus Orientation

Think tanks have considerable influence in the discourse on climate change and climate politics in Germany. However, in order to be influential, think tanks have to operate in an institutionalized environment, in which anthropogenic climate change is considered to be a scientifically proven fact. According to Weingart et al., some of the most influential think tanks such as the Potsdam Institute for Climate Impact Research and the Wuppertal Institute for Climate, Environment and Energy were purpose built to provide policy advice under the condition of global climate change (Weingart et al. 2000: 269). Since the German Reunification a system of policy-related research and advice has been institutionalized, with formalized, publicly founded networks such as the Leibniz Association and the Helmholtz Association providing crucial research to policymakers and maintaining close ties to public authorities.

The space for think tanks in general and advocacy think tanks in particular is therefore limited (McGann and Weaver 2009: 117). Moreover, since corporations 'have abstained from major interventions in the German debate and did *not* [emphasis added, AR] position themselves as climate change skeptics' (Schäfer 2016: 11) (very limited), private funding also favours advocacy that conforms with the overall consensus on climate change.

Thurid Hustedt describes the political climate for climate politics in Germany as follows:

> An analysis of the parliamentary discourses on climate change and climate policy reveals that it has been strongly related to natural-science based knowledge and has been framed as a natural-science based problem depending upon scientific research findings and their credibility
>
> (...). This problem frame has become manifest in launching large-scale research programs by federal and some Länder governments (...) and in establishing specialized research institutes such as the Potsdam Institute for Climate Impact Research (PIK) and the Wuppertal Institute for Climate, Environment and Energy in the early 1990s (...) to increase the knowledge base on anthropogenic climate change. (2013: 97)

German think tanks operate in a highly structured and stratified institutional environment. Formal membership to scholarly association, adherence to the standards of good scientific practice, and close ties to German universities are a prerequisite for getting access to public funding. This implies that the scope and the content of advice that can resonate with political authorities is also limited. As again Hustedt points out, external advisers (including think tanks and political foundations) 'provide rather long-term advice and serve a consensus-building function i.e. process uncertain and potentially contentious scientific knowledge in a consistent manner into reports and recommendations delivered to the respective lead ministries' (Hustedt 2013: 104).

As a consequence of this highly institutionalized environment, a relative small number of think tanks occupy privileged positions. The Potsdam Institute for Climate Impact Research; the Wuppertal Institute for Climate, Environment and Energy; and the Öko-Institute are key advisors on climate politics in the Federal Republic.

The previous two chapters dealt with the particularities of think tanks in the United States (Chap. 4) and Germany (Chap. 5), respectively. In the following, final chapter a systematic comparison of the two countries will focus on structural equivalents and major differences to further improve the understanding the strategic activities of think tanks.

Appendix 1

Table 5.2 Environmental think tanks | SNA data

Name[a]	*Type*	*Sum line values*[b]	*Closeness centrality*
Adelphi[a]	Advocacy	1	
Federal Ministry[a] of Economics and Technology	Government	1	
ZEW	Academic	5	0.2855850
European Union	Government	30	0.3499820
Federal Ministry of Education and Research	Government	31	0.3838510
Pakt für Forschung und Innovation	Government	1	0.2137610
CSCP[a]	Academic	2	
Henkel Corporation[a]	Industry	1	
UNEP[a]	Government	2	
IWF Kiel	Academic	6	0.2878880
Fraunhofer Institute	Academic	3	0.2531780
KIT	Academic	1	0.2150490
MCC Berlin[a]	Academic	3	
IIASA[a]	Advocacy	1	
EDF[a]	Advocacy	1	

(*continued*)

Table 5.2 (continued)

Name[a]	*Type*	*Sum line values*[b]	*Closeness centrality*
MCII	Advocacy	5	0.2461940
Federal Ministry of the Environment	Government	5	0.3216050
Munich Re	Industry	1	0.1908990
Germanwatch	Advocacy	1	0.1908990
ICCCAD	Academic	1	0.1908990
UNU-EHS	Academic		0.1908990
Öko-Institute	Advocacy	4	0.2950260
ISOE	Academic	4	0.2833180
PIK	Academic	90	0.4353430
Climate KIC	Advocacy	4	0.2878880
German Aerospace Center	Academic	16	0.2878880
GIZ	Government	1	0.2878880
German Research Foundation	Government	5	0.2878880
Project Management Jülich	Academic	7	0.2878880
Kurt Lange Foundation	Advocacy	2	0.2878880
Volkswagen Foundation	Industry	1	0.2878880
Federal Environment Agency	Government	8	0.3642660
Federal Ministry for Economic Affairs and Energy	Government	1	0.2878880
European Institute for Innovation and Technology	Academic	1	0.2878880
Deutsche Bundesstiftung Umwelt	Government	1	0.2878880
Federal Ministry of Food and Agriculture	Government	1	0.2878880
World Bank	Government	1	0.2878880
Helmholtz Centre for Environmental Research	Academic	2	0.2496370
CESifo	Advocacy	1	0.2644300
Ecologic	Academic	3	0.2644300
Climate Media Factory	Advocacy	1	0.2479040
e-fect consulting	Advocacy	1	0.2479040
Federal Institute for Research on Building, Urban Affairs and Spatial Development	Academic	1	0.2479040
Wuppertal Institute	Academic	5	0.3718550
Mercator Foundation	Industry	1	0.2586820
DIW	Academic	7	0.3305380
Leibniz Institute of Ecological Urban and Regional Development	Academic	3	0.2833180
Federal Agency for Nature Conservation	Government	1	0.2124890
IÖW	Academic	9	0.3499820
Ecofys	Advocacy	1	0.2379870
Environmental Policy Research Centre	Academic	3	0.2788920

[a]Actors marked with an asterisk are not part of the main component, closeness centrality measure not applicable

[b]"Sum of line values" gives a measure of the overall connectivity of the respective actor

Appendix 2

Table 5.3 List of think tanks in Germany

Name	*Type*	*Staff*	*Public/ private*	*Founded*	*Output*
Adelphi Research	Academic/ mixed	80	Private	2001	Books, scientific papers, contract research, newsletters, interviews, newspaper articles, conferences, and seminars
Akademie für Raumforschung und Landesplanung	Academic	190	Public	1946	Books, scientific papers, conferences, and seminars
Arbeitsgemeinschaft für Friedens- und Konfliktforschung	Academic	–	Private	1968	Books, newsletters, conferences
Arbeitsgemeinschaft für Wirtschaftliche Verwaltung	Academic	7	Mixed	1926	Books, newsletters, seminars
Arbeitsgemeinschaft Kriegsursachenforschung	Academic	12	Mixed	1986	Books, scientific papers
Arnold-Bergstraesser-Institut für kulturwissenschaftliche Forschung e.V. (ABI)	Academic	30	Mixed	1959	Books, scientific papers, conferences, and seminars
Aspen Institute, Berlin	Advocacy	11	Mixed	1974	Books, newsletter interviews, newspaper articles, conferences, and seminars
Berlin Information Center for Transatlantic Security (BITS)	Academic	8	Private	1991	Books, scientific papers, policy briefs, newspaper articles
Berlin-Institut für Bevölkerung und Entwicklung	Academic	4	Private	2000	Books, scientific papers, newsletters, newspaper articles
Bertelsmann Foundation	Academic	144	Private	1977	Books, newsletters, interviews, newspaper articles, conferences, and seminars
Bonn International Center for Conversion gGmbH	Academic	32	Public	1994	Books, scientific papers, contract research, policy brief, newsletters, conferences, and seminars
Brandenburger-Berliner Institut für Sozialwissenschaftliche Studien e.V. (BISS)	Academic	5	Private	1990	Books, scientific papers

(*continued*)

Table 5.3 (continued)

Name	*Type*	*Staff*	*Public/ private*	*Founded*	*Output*
Brandenburgisches Institut für Gesellschaft und Sicherheit gGmbH	Academic	5	Mixed	2008	Scientific papers
Centrum für angewandte Politikforschung (CAP)	Academic	41	Mixed	1995	Books, scientific papers, policy briefs, newsletters, interviews, newspaper articles, conferences, and seminars
Centrum für Europäische Politik (CEP)	Academic	6	Private	2006	Policy briefs, newsletters
Centrum für Hochschulentwicklung (CHE)	Advocacy	17	Private	1994	Books, scientific papers, newsletters, newspaper articles
Club of Rome Germany	Academic	4	Private	1968	Books, scientific papers, newsletters, interviews, newspaper articles, conferences
Council on Public Policy	Advocacy	–	Private	2001	Books, newsletters, newspaper articles, conferences, and seminars
Democracy Reporting International	Advocacy	6	Private	2006	Contract research, policy briefs, newsletters, seminars
Deutsche Gesellschaft für Auswärtige Politik e.V.	Academic	21	Mixed	1955	Books, scientific papers, contract research, policy briefs, newspaper articles, conferences, and seminars
Deutsche Gesellschaft für Umwelterziehung e.V.	Advocacy	9	Private	1983	Contract research, newsletters, conferences, and seminars
Deutsches Digital Institut	Academic	–	Private	2006	Books, contract research, conferences, and seminars
Deutsches Forschungsinstitut für Öffentliche Verwaltung (FÖV)	Academic	33	Public	1976	Books, scientific papers, newsletters, conferences
Deutsches Institut für Altersvorsorge	Advocacy	–	Private	1997	Books, contract research

(*continued*)

Table 5.3 (continued)

Name	*Type*	*Staff*	*Public/ private*	*Founded*	*Output*
Deutsches Institut für Entwicklungspolitik	Academic	103	Public	1964	Books, scientific papers, contract research, policy briefs, newsletter, interviews, newspaper articles, conferences, and seminars
Deutsches Institut für Erwachsenenbildung	Academic	67	Mixed	1957	Books, scientific papers, newsletters, conferences, and seminars
Deutsches Institut für Internationale Pädagogische Forschung	Academic	164	Mixed	1951	Books, scientific papers, newsletters, conferences
Deutsches Institut für kleine und mittlere Unternehmen e. V. (DIKMU)	Academic	20	Private	2001	Books, newsletters
Deutsches Institut für Menschenrechte	Advocacy	30	Public	2001	Books, scientific papers, contract research, policy briefs, newsletter, interviews, newspaper articles, conferences, and seminars
Deutsches Institut für Urbanistik	Academic	110	Mixed	1973	Books, scientific papers, contract research, policy briefs, newsletters, interviews, newspaper articles, conferences, and seminars
Deutsches Institut für Wirtschaftsforschung (DIW)	Academic	165	Mixed	1926	Books, scientific papers, contract research, policy briefs, newsletters, interviews, newspaper articles, conferences, and seminars
Deutsches Jugendinstitut	Academic	140	Public	1963	Books, scientific papers, newsletters, conferences
Deutsch-Französisches Institut	Advocacy	25	Mixed	1948	Books, scientific papers, policy briefs, newsletters, conferences, and seminars

(*continued*)

Table 5.3 (continued)

Name	*Type*	*Staff*	*Public/ private*	*Founded*	*Output*
Düsseldorfer Institut für Außen- und Sicherheitspolitik e.V.	Academic	33	Private	2003	Books, policy briefs, newsletters interviews, newspaper articles
Ecologic	Academic	101	Private	1995	Books, scientific papers, contract research newsletters, newspaper articles, conferences
Econwatch	Academic	–	Private	2007	Policy briefs, newsletters, conferences, and seminars
Eduard Pestel Institut für Systemforschung e.V.	Academic	3	Private	1975	Contract research
Europäische Akademie zur Erforschung von Folgen wissenschaftlich-technischer Entwicklungen	Academic	20	Mixed	1996	Books, scientific papers, newsletters, conferences
Europäisches Forum für Migrationsstudien e.V. (efms)	Academic	10	Mixed*	1993	Books, scientific papers, contract research, conferences, and seminars
Europäisches Migrationszentrum (EMZ)	Academic	–	–	1978	Books, scientific papers, contract research, conferences
Europäisches Zentrum für Minderheitenfragen (ECMI)	Academic	16	Public	1996	Books, scientific papers, newsletters, conferences, and seminars
European Committee for a Constructive Tomorrow (CFACT)	Advocacy*	–	Private	2004	Books, scientific papers, contract research, newsletters, newspaper articles, conferences
European Council on Foreign Relations	Advocacy	15	Private	2007	Books, policy briefs, newsletter, interviews, newspaper articles
European Stability Initiative (ESI)	Academic	16	Private	1999	Scientific papers, contract research, newsletter, interviews, newspaper articles, conferences, and seminars
Finanzwissenschaftliches Institut an der Universität zu Köln e.V.	Academic	11	Private	1927	Books, scientific papers, contract research, newsletter

(*continued*)

Table 5.3 (continued)

Name	*Type*	*Staff*	*Public/ private*	*Founded*	*Output*
Forschungsinstitut für Bildungs- und Sozialökonomie	Academic	17	Private	1993	Books, scientific papers, contract research, newspaper articles, conferences, and seminars
Forschungsinstitut für Ordnungspolitik gGmbH (FiO)	Academic	–	Mixed*	1958*	Books, newspaper articles
Forschungsinstitut für Philosophie Hannover (FIPH)	Academic	12	Private	1988	Books, conferences
Forschungsinstitut zur Zukunft der Arbeit (IZA)	Academic	43	Private	1998	Books, scientific papers, policy briefs, contract research, newsletter, interviews, newspaper articles, conferences, and seminars
Forschungsstätte der Evangelischen Studiengemeinschaft (FEST)	Advocacy	23	Private	1958*	Books, scientific papers, newsletters
Forschungsstelle Osteuropa	Academic	34	Public	1982	Books, scientific papers, policy briefs, conferences
Forum Ökologisch-Soziale Marktwirtschaft	Advocacy	9	Private	1994	Scientific papers, contract research, newsletters, interviews, newspaper articles, conferences
Friedrich-Ebert-Foundation (FES)	Advocacy	571	Public	1925	Books, scientific papers, contract research, newsletters, interviews, newspaper articles, conferences, and seminars
Friedrich-Naumann-Foundation	Advocacy	345	Public	1958	Books, policy briefs, conferences, and seminars
GIGA German Institute of Global and Area Studies	Academic	50	Public	1964	Books, scientific papers, contract research, policy briefs, interviews, newspaper articles, conferences
Global Public Policy Institute	Academic	18	Private	2003	Books, scientific papers, contract research, newsletters, newspaper articles, seminars

(*continued*)

Table 5.3 (continued)

Name	*Type*	*Staff*	*Public/ private*	*Founded*	*Output*
Hamburger Institut für Sozialforschung (HIS)	Academic	72	Private*	1984	Books, scientific papers, conferences
Hamburger Umweltinstitut-Zentrum für soziale und ökologische Technik	Academic	10	Private	1989	Contract research
Hamburgisches Weltwirtschaftsinstitut (HWWI)	Academic	64	Private	2005	Books, scientific papers, contract research, policy briefs, newsletters, interviews, newspaper articles, conferences, and seminars
Hanns Seidel Foundation	Advocacy	660	Public	1967	Books, newsletters, conferences, and seminars
Haus Rissen – Internationales Institut für Politik und Wirtschaft	Academic	11	Private	1954	Books, scientific papers, policy briefs, newsletter, interviews, newspaper articles, conferences, and seminars
Heinrich-Böll-Foundation	Advocacy	168	Public	1997	Books, scientific papers, newsletters, newspaper articles, conferences, and seminars
Hessische Stiftung Friedens- und Konfliktforschung	Academic	60	Public	1970	Books, scientific papers, newsletters, interviews, newspaper articles, conferences
IFEU* (not included in TTD)	Advocacy	60	Private	1978	Books, scientific papers, contract research
Ifo Institut für Wirtschaftsforschung	Academic	170	Mixed	1949	Books, scientific papers, contract research, policy briefs, newsletters, interviews, newspaper articles, conferences
Institut Arbeit und Technik	Academic	50	Public	1988	Books, scientific papers, contract research, newsletters, interviews, newspaper articles, conferences

(*continued*)

Table 5.3 (continued)

Name	*Type*	*Staff*	*Public/ private*	*Founded*	*Output*
Institut Arbeit und Qualifikation	Academic	54	Public	2007	Books, scientific papers, newsletters, interview, newspaper articles, conferences, and seminars
Institut der deutschen Wirtschaft Köln	Advocacy	80*	Private*	1951*	Books, contract research, policy briefs, newsletters, interviews, newspaper articles, conferences, and seminars
Institut für Allgemeine Übersee-Forschung	Academic	9	Public	1964	Books, scientific papers
Institut für Angewandte Familien-, Kindheits- und Jugendforschung (IFK)	Academic	14	Mixed	1990	Books, scientific papers, contract research, conferences
Institut für angewandte Verkehrs- und Tourismusforschung	Academic	–	Private	1984	Contract research
Institut für Angewandte Wirtschaftsforschung (IAW)	Academic	13	Mixed	1957	Books, scientific papers, contract research, newsletters, conferences, and seminars
Institut für Arbeitsmarkt- und Berufsforschung der Bundesanstalt für Arbeit (IAB)	Academic	343*	Public	1967	Books, scientific papers, policy briefs, newsletters, interviews, newspaper articles, conferences
Institut für christliche Ethik und Politik	Academic	4	Mixed	2004	Contract research, policy briefs, conferences
Institut für Entwicklungsplanung und Strukturforschung	Academic	12	Private	1972	Books, scientific papers, contract research
Institut für Europäische Politik e. V. (IEP)	Academic	15	Mixed*	1959	Books, scientific papers, newsletters, conferences, and seminars
Institut für Friedensforschung und Sicherheitspolitik (IFSH)	Academic	60	Mixed	1971	Books, scientific papers, contract research, interviews, newspaper articles, conferences, and seminars
Institut für Gesundheits- und Sozialforschung	Academic	30	Private	1980	Books, scientific papers, contract research, conferences

(*continued*)

Table 5.3 (continued)

Name	*Type*	*Staff*	*Public/ private*	*Founded*	*Output*
Institut für Länderkunde e.V. (IfL)	Academic	60	Public	1992	Books, scientific papers, interviews, newspaper articles, conferences
Institut für Landes- und Stadtentwicklungsforschung (ILS)	Academic	250	Public	2003	Books, contract research, newsletters
Institut für Makroökonomie und Konjunkturforschung	Advocacy	15	Public (Hans-Böckler Foundattion)	2005	Books, scientific papers, contract research, policy briefs, newsletters, interviews, newspaper articles, conferences
Institut für Markt – Umwelt – Gesellschaft e.V. (IMUG)	Academic	20	Mixed*	1992	Books, scientific papers, contract research, newsletters
Institut für Medien- und Kommunikationspolitik (IfM)	Academic	5	Mixed	2005	Books, scientific papers, contract research, newspaper articles, seminars
Institut für Mittelstandsforschung Bonn (IfM)	Academic	22	Public	1957	Books, scientific papers, newsletters
Institut für Mobilitätsforschung	Academic	6	Private	1998*	Books
Institut für Ökologische Wirtschaftsforschung	Academic	30	Mixed	1985	Books, scientific papers, contract research, conferences
Institut für Sozialforschung und Gesellschaftspolitik e.V. (ISG) Otto-Blume Institut	Academic	23*	Private	1952	Contract research
Institut für sozial-ökologische Forschung	Academic	27	Mixed	1989	Books, scientific papers, contract research, newsletters, newspaper articles, conferences
Institut für sozialökonomische Strukturanalysen Berlin (SÖSTRA)	Academic	14	Private	1990	Contract research
Institut für Sozialwissenschaftliche Analysen und Beratung (ISAB)	Academic	7*	Private*	1987	Books, contract research, newsletters, conferences, and seminars
Institut für Sozialwissenschaftliche Forschung e.V. (ISF)	Academic	30	Private	1965	Books, scientific papers, contract research
Institut für Strukturpolitik und Wirtschaftsförderung Halle-Leipzig e.V. (ISW)	Academic	80*	Mixed*	1991	Books, contract research, conferences

(*continued*)

Table 5.3 (continued)

Name	*Type*	*Staff*	*Public/ private*	*Founded*	*Output*
Institut für Technikfolgenabschätzung und Systemanalyse (ITAS)	Academic	60	Public	1995	Books, scientific papers, conferences
Institut für Terrorismusforschung und Sicherheitspolitik (IFTUS)	Academic	–	Private*	2003	Contract research, newsletters, interviews, newspaper articles, conferences, and seminars
Institut für Unternehmerische Freiheit e. V. (iuf)	Advocacy	7	Private	1998	Scientific papers, contract research, newsletters, interviews, newspaper articles, conferences, and seminars
Institut für Weltwirtschaft (IfW)	Academic	270	Public	1914	Books, scientific papers, conferences
Institut für Wirtschaftsforschung Halle	Academic	79	Public	1992	Books, scientific papers, contract research, policy briefs, newsletters, newspaper articles
Institut für Wirtschaftspolitik an der Universität Köln	Academic	8	Mixed*	1950	Books, scientific papers, policy briefs
Institut für Wissenschaft und Ethik	Academic	18	Public*	1993	Books, scientific papers, contract research, newsletters, conferences
Institut für Wirtschaft und Gesellschaft Bonn e.V.	Academic	4	Public*	1977	Books, scientific papers, contract research, newspaper articles
Institut für Zukunftsstudien und Technologiebewertung	Academic	36*	Mixed*	1981	Books, scientific papers, contract research, newsletters, interviews, conferences
Institut Solidarische Moderne	Advocacy	–	Private	2010	Newsletters, interviews, newspaper articles, conferences
Institut des Bundes der Steuerzahler*	Advocacy		Private	1965*	Books, newspaper articles*
KATALYSE Institut für angewandte Umweltforschung	Academic	17	Private	1978	Books, contract research, newspaper articles
Konrad-Adenauer-Foundation	Advocacy	527	Mixed	1956	Books, scientific papers, policy briefs, newsletters, conferences, and seminars

(*continued*)

Table 5.3 (continued)

Name	*Type*	*Staff*	*Public/ private*	*Founded*	*Output*
Leibniz-Institut für Agrarentwicklung in Mittel- und Osteuropa (IAMO)	Academic	77	Public	1994	Books, scientific papers, newsletters, conferences, and seminars
Leibniz-Institut für Ökologische Raumentwicklung e.V.	Academic	105	Public	1992	Books, scientific papers, newsletters, interviews, conferences
Leibniz-Institut für Regionalentwicklung und Strukturplanung e.V. (IRS)	Academic	88	Public	1992	Books, scientific papers, newsletters, conferences
Mannheimer Forschungsinstitut Ökonomie und Demographischer Wandel	Academic	22	Mixed	2001	Books, scientific papers, policy briefs, newsletters, conferences
Max Planck Institute for Research on Collective Goods	Academic	20	Mixed*	2003	Books, scientific papers
Max-Planck-Institut für Gesellschaftsforschung	Academic	57	Mixed	1985	Books, scientific papers, newsletters, newspaper articles, conferences
Mittelstandsinstitut Niedersachsen e.V.	Advocacy	–	Private*	1975	Books, scientific papers*
Niedersächsisches Institut für Wirtschaftsforschung	Academic	14	Mixed	1981	Books, contract research, newsletters, seminars
Öko-Institut	Advocacy	130	Private	1977	Books, scientific papers, contract research, newsletters, newspaper articles, conferences, and seminars
Open Europe Berlin	Advocacy	5	Private	2012	Newsletters, interviews, newspaper articles
Osteuropa-Institut	Academic	33	Mixed	1952	Books, scientific papers, contract research, policy briefs, newsletters, conferences
Oswald-von-Nell-Breuning Institut für Wirtschafts- und Gesellschaftsethik	Advocacy	4	Mixed*	1990*	Books, interviews*, newspaper articles, conferences
PlanBplus	Academic	6	Private	2006	Conferences
Potsdam-Institut für Klimafolgenforschung (PIK)	Academic	137	Mixed	1992	Books, scientific papers, newspaper articles, interviews*, conferences

(*continued*)

Table 5.3 (continued)

Name	*Type*	*Staff*	*Public/ private*	*Founded*	*Output*
Progressives Zentrum e. V.	Advocacy	8	Private	2007	Policy briefs, newsletters, interviews, newspaper articles, conferences
Rat für Migration e.V.	Academic	150*	Private*	1998	Books, interviews, newspaper articles
Rheinisch-Westfälisches Institut für Wirtschaftsforschung	Academic	82	Public	1926	Books, scientific papers, contract research
Rhein-Ruhr-Institut für Sozialforschung und Politikberatung e.V. (RISP)	Academic	60	Private	1980	Books, contract research
Ruhr Forschungsinstitut für Innovations- und Strukturpolitik e.V. (RUFIS)	Academic	8*	Private*	1979	Books, contract research
Rosa-Luxemburg-Foundation	Advocacy	50	Public	1992	Books, scientific papers, policy briefs, newsletters
Schleswig-Holsteinisches Institut für Friedensforschung (SCHIFF)	Academic	7	Public	1995	Books, scientific papers, seminars
Sekretariat für Zukunftsforschung	Academic	9	Private	1981	Books, contract research
Sozialforschungsstelle Dortmund	Academic	50	Public	1946	Books, scientific papers, conferences, and seminars
Sozialwissenschaftliches Institut der Evangelischen Kirche in Deutschland	Advocacy	11*	Private*	2004*	Books, scientific papers, newsletters, interviews, newspaper articles, conferences, and seminars
Stiftung Entwicklung und Frieden (SEF)	Academic	4	Public	1986	Books, policy briefs, newsletters, conferences
Stiftung Marktwirtschaft	Advocacy	8	Private	1982	Policy briefs, newsletters, interviews, newspaper articles, conferences, and seminars
Stiftung neue Verantwortung	Academic	14	Private	2008	Scientific papers, policy briefs, newsletters, interviews, newspaper articles, conferences, and seminars

(*continued*)

Table 5.3 (continued)

Name	*Type*	*Staff*	*Public/ private*	*Founded*	*Output*
Stiftung Ordnungspolitik (SOP)	Advocacy	9	Mixed*	1999	Books, interviews, newspaper articles, conferences
Stiftung Wissenschaft und Politik (SWP)	Academic	110	Public	1962	Books, scientific papers, policy briefs, newsletters, interviews, newspaper articles
Studienzentrum Weikersheim e.V.	Advocacy	–	Private	1979	Books, newsletters, conferences
Trierer Arbeitsgemeinschaft für Umwelt-, Regional- und Strukturforschung e.V. an der Universität Trier	Academic	12	Private	1998	Contract research, newsletters
Umweltforschungszentrum Leipzig-Halle GmbH (UFZ)	Academic	780	Public	1991	Books scientific papers, conferences, and seminars
Umwelt und Prognose Institut (UPI)	Academic	7	–	–	Policy briefs
Unternehmerinstitut e.V. (UNI) der Arbeitsgemeinschaft Selbstständiger Unternehmer (ASU)	Advocacy	–	Private	1993	Policy briefs, conferences
Walter Eucken Institut	Academic	5	Mixed	1954	Books, scientific papers, newspaper articles, conferences, and seminars
Walter-Raymond-Stiftung der Bundesvereinigung der Deutschen Arbeitgeberverbände	Advocacy	–	Private	1959	Books, conferences
Walther Rathenau Institut – Stiftung für internationale Politik	Advocacy	–	Private	2008	Scientific papers, contract research, interviews, newspaper articles, conferences
Wirtschafts- und Sozialwissenschaftliches Institut (WSI)	Advocacy	38	Private	1946	Books, scientific papers, newsletters, newspaper articles, conferences
Wissenschaftszentrum Berlin für Sozialforschung (WZB)	Academic	357	Mixed	1969	Books, scientific papers, newsletters, interviews, newspaper articles, conferences
Wittenberg-Zentrum für globale Ethik e.V.	Academic	13*	Private*	1998	Newsletters, seminars

(*continued*)

Table 5.3 (continued)

Name	*Type*	*Staff*	*Public/ private*	*Founded*	*Output*
Wuppertal Institut für Klima, Umwelt, Energie	Academic	140	Public	1991	Books, scientific papers, contract research, newsletters, conferences
Zentrum für Entwicklungsforschung	Academic	50	Public	1995	Books, scientific papers, policy briefs, newsletters, newspaper articles, conferences, and seminars
Zentrum für Europäische Integrationsforschung	Academic	50	Public*	1995	Books, scientific papers, conferences, and seminars
Zentrum für Europäische Rechtspolitik	Academic	16	Private	1982	Books, contract research
Zentrum für Europäische Wirtschaftsforschung GmbH	Academic	181	Mixed	1990	Books, scientific papers, contract research, newsletters, interviews, newspaper articles, conferences, and seminars
Zentrum für gesellschaftlichen Fortschritt	Academic	1	Private	2009	Books, newsletters, interviews, newspaper articles
Zentrum für Kulturforschung (ZfKf)	Academic	10	Private	1970	Contract research

Note: The "Institut für anwendungsorientierte Innovations- und Zukunftsforschung" and the "Institute für demographische Zukunftsfähigkeit" despite still being listed were excluded because they were either dissolved ("Institut für anwendungsorientierte Innovations- und Zukunftsforschung") or no longer active ("Institute für demographische Zukunftsfähigkeit").

Notes

1. Founded in 1948 as the successor for the Kaiser Wilhelm Society, the Max-Planck-Society operates 83 so-called Max-Planck-Institutes to foster basic research. Although legally registered as an independent association, the Max-Planck-Society is publicly funded with only a small share stemming from third-party funding and donations (cf. https://www.mpg.de/facts-and-figures).
2. Founded in 1949, 'Fraunhofer is Europe's largest application-oriented research organization. Our research efforts are geared entirely to people's needs: health, security, communication, energy and the environment' (https://www.fraunhofer.de/en/about-fraunhofer/profile-structure.

html). The Fraunhofer Society is the largest and most important association for contracted research: 'The Fraunhofer-Gesellschaft employs a staff of 24,500, who work with an annual research budget totalling 2.1 billion euros. Of this sum, 1.9 billion euros is generated through contract research. More than 70 percent of the Fraunhofer-Gesellschaft's contract research revenue is derived from contracts with industry and from publicly financed research projects' (https://www.fraunhofer.de/en/about-fraunhofer/profile-structure/facts-and-figures.html).

3. The thinktankdirectory database (thinktankdirectory.org) serves as the starting point for compiling the sample of think tanks in Germany. The data was checked and updated in August 2017. Missing think tanks, such as the IFEU, had been added to the sample.
4. Political foundations are counted as advocacy think tanks.
5. The findings indicate an increase in the awareness of climate change as a serious problem in the US. Drawing on the 2003 *World Public Opinion Survey*, Harrison and Sundstrom found that then 31 per cent of the Americans thought of climate change as a very serious problem. With 54 per cent of Germans shared this conviction in 2003, the number has been very stable in the Federal Republic (Harrison and Sundstrom 2007: 7).
6. With exception of the Left Party (first called "PDS" since 2005 "Linkspartei" or "Die Linke"), all political parties with political foundations have been members in a government coalition on the federal level. It is therefore not surprising that the positions of political foundation do not differ and considerably form government position on climate politics.

References

Adam, Thomas. 2004. *Philanthropy, Patronage, and Civil Society. Experiences from Germany, Great Britain and North America*. Bloomington: Indiana University Press.

Brown, James R. 2001. *Who Rules in Science?* Cambridge, MA/London: Harvard University Press.

Burchard, Amory. 2015. Streit um AfD Stiftung: EU-Kommisar könnte gegen "Erasmus"-Stipendien vorgehen. *Der Tagesspiegel*, August 17. http://www.tagesspiegel.de/wissen/streit-um-afd-stiftung-eu-kommissar-koennte-gegen-erasmus-stipendien-vorgehen/12194964.html. Accessed 29 Aug 2017.

Dunn, Stephen, and John Gennard. 1984. *The Closed Shop in British Industry*. London/Basingstoke: Macmillan.

Esping-Andersen, Gosta. 1990. *The Three Worlds of Welfare Capitalism*. Princeton: Princeton University Press.

Frye, Northrop. 1974. The Decline of the West by Oswald Spengler. *Daedalus* 103 (1): 1–13.

Gemeinsame Wissenschaftskonferenz, G.W.K. 2015. *Pakt für Forschung und Innovation Monitoring Bericht 2015*. Bonn: Materialien der GWK.

Harrison, Kathryn, and Lisa McIntosh Sundstrom. 2007. The Comparative Politics of Climate Change. *Global Environmental Politics* 7 (4): 1–18.

Hopwood, Nick. 1996. Producing a Socialist Popular Science in the Weimar Republic. *History Workshop Journal* 41: 117–153.

Hustedt, Thurid. 2013. Analysing Policy Advice: The Case of Climate Policy in Germany. *Central European Journal of Public Policy* 7 (1): 88–111.

Jochem, Sven. 2013. Think Tanks: Bridging Beltway and Ivory Tower? In *Policy Analysis in Germany*, ed. Sonja Blum and Klaus Schubert, 231–246. Bristol: Policy Press.

Kaes, Anton, Martin Jay, and Edward Dimendberg. 1994. *The Weimar Republic Sourcebook*. Berkeley/Los Angeles/London: University of California Press.

Krück, Carsten, Jutta Borchers, and Peter Weingart. 1999. Climate Research and Climate Politics in Germany: Assets and Hazards of Consensus-Based Risk Management. University of Bielefeld, Department of Sociology. http://pne.people.si.umich.edu/PDF/PMNPC/germany.pdf

Mayntz, Renate. 1987. West Germany. In *Advising the Rulers*, ed. William Plowden, 3–18. Oxford: Basil Blackwell.

McGann, James, and Kent Weaver. 2009. *Think Tanks and Civil Societies: Catalysts for Ideas and Action*. New Brunswick/London: Transaction Publishers.

McPherson, Miller, Lynn Smith-Lovin, and James M. Cook. 2001. Birds of a Feather: Homophily in Social Networks. *Annual Review of Sociology* 27: 415–444.

Ott, Hermann E. 2007. Internationale Klimapolitik 2020. Herausforderung für die deutsche (Umwelt-)Außenpolitik. *Kompass 2020*. Berlin: Friedrich-Ebert-Stiftung.

Pautz, Hartwig. 2010. Think Tanks in the United Kingdom and Germany: Actors in the Modernisation of Social Democracy. *The British Journal of Politics and International Relations* 12: 274–294.

PEW Research Centre. 2015. *Global Concern About Climate Change, Broad Support for Limiting Emissions*. November 5.

Pflanze, Otto. 1968. *Bismarck and the Development of Germany. The Period of Unification 1815–1871*. Princeton: Princeton University Press.

Pogorelskaja, Swetlana W. 2002. Die parteinahen Stiftungen als Akteure und Instrumente der deutschen Außenpolitik. *Aus Politik und Zeitgeschichte* 6–7: 29–38.

Rufloff, Wilfried. 2004. Wissenschaftliche Politikberatung in der Bundesrepublik—historische Perspektive. In *Politikberatung in Deutschland. Praxis und Perspektiven*, ed. Steffen Dagger, Christoph Greiner, and Kirsten Leinert, 178–199. Wiesbaden: VS-Verlag für Sozialwissenschaften.

Schäfer, Mike S. 2016. Climate Change Communication in Germany. *Oxford Research Encyclopedia of Climate Science*: 1–19.

Speth, Rudolf. 2010. Stiftungen und Think Tanks. In *Handbuch Wissenschaftspolitik*, ed. Dagmar Simon, Andreas Knie, and Stefan Hornbostel, 390–405. Heidelberg: Springer.

Stone, Diane. 2007. Recycling Bins, Garbage Cans or Think Tanks? Three Myths Regarding Policy Analysis Institute. *Public Administration* 85 (2): 259–278.

Thunert, Martin. 2001. Politikberatung in der Bundesrepublik Deutschland seit 1949. In *Demokratie und Politik in der Bundesrepublik 1949–1999*, ed. Ulrich Willems, 223–242. Wiesbaden: VS-Verlag für Sozialwissenschaften.

———. 2004. Think Tanks in Germany. In *Think Tank Traditions. Policy Research and the Politics of Ideas*, ed. Diane Stone and Andrew Denham, 71–88. Manchester/New York: Manchester University Press.

von Arnim, Hans Herbert. 1991. *Die Partei, der Abgeordnete und das Geld*. Mainz: Hase & Koehler Publishing.

von Vieregge, Henning. 1977. »Globalzuschüsse« für die parteinahen Stiftungen: Parteienfinanzierung auf Umwegen? *Zeitschrift für Parlamentsfragen* 8 (1): 51–58.

———. 1990. Die Partei-Stiftungen: Ihre Rolle im politischen System. In *Parteienfinanzierung und politischer Wettbewerb*, ed. Göttrik Werner, 164–194. Opladen: Westdeutscher Verlag.

von Winter, Thomas. 2006. Die Wissenschaftlichen Dienste des Deutschen Bundestages. In *Handbuch Politikberatung*, ed. Svenja Falck, Dieter Rehfeld, Andrea Römmele, and Martin Thunert, 198–214. Wiesbaden: VS-Verlag für Sozialwissenschaften.

Weingart, Peter, Anita Engels, and Petra Pansegrau. 2000. Risks of Communication Discourses on Climate Change in Science, Politics, and the Mass Media. *Public Understanding of Science* 9: 261–283.

CHAPTER 6

German and US Think Tanks in Comparison

With all of the hysteria, all of the fear, all of the phony science, could it be that manmade global warming is the greatest hoax ever perpetrated on the American people?
Senator James Inhofe "The Science of Climate Change", Senate Floor Speech 28 July 2003

Introduction

The (in his view, rhetorical) question raised by US senator and renowned climate sceptic James Inhofe would draw distinctively different responses in the United States and Germany. In Germany, the most likely answer would be a "no". No, global warming isn't a hoax. And it isn't a hoax because science tells us that it's real. In the US, however, the response would depend on the political conviction of the respondent. The reason for this different approach to climate change, as we have seen, isn't that science differs in the respective countries. The reason is that the complicated findings of climate science are translated and debated differently.

The two preceding chapters described think tanks in their respective environment. It was shown that in both countries, the most visible think tanks are embedded in specific social networks. In the United States, conservative groups which share a belief system that circles around radical free-market ideas and a hostility towards government intervention provide the most important context of think tank engaged in debates on climate

A. Ruser, *Climate Politics and the Impact of Think Tanks*,
https://doi.org/10.1007/978-3-319-75750-6_6

change and climate politics: Supported by wealthy donors, conservative think tanks like the Competitive Enterprise Institute, the Heartland Institute, and the CATO Institute challenge climate politics by providing expertise to like-minded policymakers and by directly addressing the wider public. In contrast, German think tanks operate in a more formally structured environment, which values formal membership in state-sponsored networks and scientific reputation.

In this final chapter these diverse findings will be systematically contrasted and compared to arrive at a better understanding of country-specific strategies and specific patterns of think tank influence. Moreover, comparing the two countries can not only contribute to a better insight of the influence of think tanks of national climate politics but also to a better theoretical understanding of think tank activities.

Since think tanks have long been regarded as a particularity of the US American or Anglo-Saxon political systems (Ricci 1993), it stands to reason that American think tanks are depicted as "role models" for think tanks elsewhere. However, the findings of this study underline that think tanks in United States and Germany differ, fill distinct roles, have diverse organizational structures, pursue specific strategies, and target different audiences.

The findings presented in the previous chapters especially cast some doubt whether the "global spread" of think tanks (McGann 2010: 19) is indicating a "catch-up development" or even a convergence political advisory. Despite a numerical increase, since the 1990s, think tanks seem to be distinctively different and pursue diverging strategies for exercising influence in the Germany.

The Impact of Knowledge Regimes: Competitive Markets of Ideas vs. Consensus-Oriented Statist Decision-Making

As outlined in the third chapter, a key aspect of understanding and explaining diverging strategies of think tanks is the respective "environment" think tanks operating in. As was further explained, the "knowledge regime" approach of John Campbell and Over Pedersen (2014) provides a suitable model for conceptualizing the respective context of think tanks.

'The U.S. knowledge regime is often described as a highly competitive marketplace of ideas. Advocates of various policy ideas battle each other in order to influence the thinking of national policymakers' write Campbell and Pedersen (2014: 39), thereby explaining why the services of advocacy think tanks are in high demand. In an environment where rivalling policy ideas are vigorously lobbied think tanks are particularly important for influencing political debate that circle around scientific knowledge. For there is no reason, that scientific knowledge claims should be excluded from the "battle" for political influence. Moreover, Campbell and Pedersen observe three developments in the United States that are crucial for understanding and explaining think tank behaviour: First, they find that the metaphor of the "marketplace of ideas" does not imply that advocates necessarily act as *individual* market players but that there is 'a significant amount of cooperation among research organizations in Washington' (ibid.). Second, they observed that "considerable blurring had occurred" (Campbell and Pedersen 2014: 40) between scholarly and advocacy organizations. Finally, they were surprised to discover that, in contrast to what their regime typology implies, 'American *state* [emphasize added, AR] research organizations are often large and well- resourced and have substantial and sophisticated analytical capabilities' (ibid.).

The findings presented in the fourth chapter are consistent with Campbell and Pedersen's observation of a blurring of the boundaries between scholarly and advocacy organizations. The authors remark that this blurring works both ways. Applied to climate politics this means that not only academic think tanks but also climate scientists themselves feel increasingly forced to adopt the communication strategies of advocacy organizations to actively support political positions in line with the political implications derived from their scientific research.

For instance, on 28 June 2016, 31 nonpartisan scientific societies, among them the *American Association for the Advancement of Science*, wrote a letter to members of congress to reaffirm the scientific consensus on climate change and demand bold political action[1]:

> *Dear Members of Congress,*
>
> *We, as leaders of major scientific organizations, write to remind you of the consensus scientific view of climate change.*
>
> *Observations throughout the world make it clear that climate change is occurring, and rigorous scientific research concludes that the greenhouse gases emitted by human activities are the primary driver. This conclusion is based on multiple independent lines of evidence and the vast body of peer-reviewed science.*

> *There is strong evidence that ongoing climate change is having broad negative impacts on society, including the global economy, natural resources, and human health. For the United States, climate change impacts include greater threats of extreme weather events, sea level rise, and increased risk of regional water scarcity, heat waves, wildfires, and the disturbance of biological systems. The severity of climate change impacts is increasing and is expected to increase substantially in the coming decades.1*
>
> *To reduce the risk of the most severe impacts of climate change, greenhouse gas emissions must be substantially reduced. In addition, adaptation is necessary to address unavoidable consequences for human health and safety, food security, water availability, and national security, among others.*
>
> *We, in the scientific community, are prepared to work with you on the scientific issues important to your deliberations as you seek to address the challenges of our changing climate.* (https://www.aaas.org/sites/default/files/06282016.pdf, accessed 29 October 2017)

Although the letter sends a clear signal and uses unequivocal language, its impact might be limited: 'I hate to sound like a wet blanket, but this is nice. It's well-intentioned, but it won't do anything' said Jon Foley, executive director of the California Academy of Sciences adding that '[w]e're being bad scientists—not in how we look at our climate data but in how we look at our communication data' (Smith 2016). Foley's comments express the increasingly widespread conviction that 'scientists are wrong to hope that simply explaining the science again will change the minds of politicians who have not listened before' (ibid.). Instead individual scientists and scientific societies should actively engage in public debates and the policymaking process.

However, this increasing engagement with the political consequences of climate research has serious consequences for climate science itself: Firstly, by deliberately blurring the boundaries between climate research and climate politics, scientists risk to strengthen the perception that they are taking political sides rather than providing an impartial scientific perspective. Secondly by becoming actively engaged in the climate politics, climate scientists enter the arena of organizations with a track record in optimizing their communication data: advocacy think tanks.

Unwittingly climate scientists thus make it easier for advocacy think tanks to blurring the boundaries between scholarly research and advocacy. As discussed in the fourth chapter, advocacy think tanks such as the Heartland Institute are actively seeking to creating a semblance of scientific objectivity and, in consequence, scientific authority. The founding of the Nongovernmental International Panel on Climate Change (NIPCC) is an attempt to imitate the procedures and structures of institutionalized "mainstream" climate science. Although the reports compiled by the NIPCC (under the auspices of the Heartland Institute) fail to meet the scientific standards of the comprehensive assessment reports of the Intergovernmental Panel on Climate Change (IPCC), they are still able to contest scientific consensus and spur further debate on the credibility and reliability of climate science (see Machin and Ruser 2018). In contrast to classical lobbying in the US which focuses on formulating and promoting policy ideas that 'tend to be technical, very specific, and tuned to the narrow interests of individual paying clients' (Campbell and Pedersen 2014: 50), the blurring of the boundaries allows advocacy think tanks to influence public sentiment (e.g. by calling into question the scientific consensus on climate change) and to formulate "political paradigms", that is, '[i]deas as elite assumptions that constrain the cognitive range of useful solutions available to policy makers' (Campbell 1998: 385). The blurring of the boundaries between scholarly research or scientific expertise and advocacy can be understood as a relocation of the main theatre of war in the battle of ideas. Applying John Campbell's model of the place of policy ideas in political discourses (see Table 3.2), we can observe an emphasis of the "background" of discourses on climate politics in the US.

Conservative think tanks have been instrumental in shaping political paradigms and influencing the public sentiment. It is likely that academic think tanks and more traditional research institutions such as universities would focus on the "foreground" of policy debate. "Supporting political "programmes" that is "the charting of "clear and specific course of policy action" (Campbell 1998: 385) and assisting in the formulation of "frames" by identifying 'symbols and concepts that help policy makers to legitimize policy solutions to the public' (ibid.) suits the methodical approach of scientific research, based on certain, transparent scientific assumptions better since it allows the researcher to 'focus on issues that can be resolved by science', that is, to play the role of a "science arbiter" (Pielke 2007: 15).

That scholars and representatives of scientific societies mourn their poor public relation is indicating that dissent on climate politics in the US is *not* stemming from scientific uncertainty. That scientific societies and individual scientist such as James Hansen (see Introduction) increasingly turn to influencing the public sentiment, and political paradigms strengthen the impression that public and political debate in climate change was successfully shifted away from scientific findings to personal, normative, political convictions: Climate science is challenged to reject environmental regulation, decry interference with free enterprise, and condemn climate politics as un-American.

In sum, scholarly organizations are facing a dilemma: Despite their best efforts and mounting evidence that supports key assumptions of climate science, political debate in the United States is nowhere close to accepting the "fact" of human-made climate change. If presenting more or "better" evidence will presumably do nothing to change this situation, scientists may abandon scientific detachment and seek to provide normative interpretations themselves.

However, since debate on climate politics takes place in an increasingly polarized environment, this strategy could play into the hands of advocacy institutions, which seek to turn the debate over climate change into a dispute over norms and values. This continuing and increasing polarization of the political landscape also affects the role and credibility of state research organizations in general and state-sponsored research on climate change in particular.

For instance, the Environmental Protection Agency (EPA), established in 1970 'to consolidate in one agency a variety of federal research, monitoring, standard-setting and enforcement activities to ensure environmental protection' (https://www.epa.gov/history), finds itself under attack from the respective opposition party. During the Obama Administration, the EPA was criticized by Republicans for supporting the presidents' "war on coal" (Davenport and Baker 2014). The criticism came after an earlier attempt to strip the EPA of most of its regulative powers by introducing the "Energy Tax Prevention Act" (Goldenberg 2011).

If Republicans saw the EPA as an extension of the Obama Administration, so do Democrats since the inauguration of Donald Trump. Even before Trump took office, Evan Lehmann and Benjamin Storrow asked in the "Scientific American" whether Democrats can "block Trump's EPA Nominee" (Lehmann and Storrow 2016) when it was only rumoured that the incoming administration might pick Scott Pruitt, a renowned climate denier who had sued the EPA several times.

As an agency of the federal government, the EPA is clearly regarded to be a political actor which might use its research capacities and administrative authorities to support the political agenda of the respective government in office.[2] Despite its "substantial and sophisticated analytical capabilities" (Campbell and Pedersen 2014: 40), the EPA is not occupying a special position.

On a marketplace of ideas, these ideas must survive competition, they must "sell". American advocacy think tanks have extensive experience in selling their message on climate change, making them perfect salesmen of a conservative agenda and sought-after manufacturers of counter-evidence.

Politicized Public Opinion on Climate Change

The metaphor of the marketplace helps understanding why ideas and accompanying "sales plans" that support a political agenda are in high demand in the US. However, it has yet to be answered why dissenting views on climate change are top sellers in the States and shelf warmers in the Federal Republic of Germany.

A first indicator is dissemination of climate sceptic views across the political spectrum in the two countries. Twenty years after the climate summit in Kyoto and almost three decades after the IPCC published its first assessment report,[3] climate change remains a contested issue. Despite the mounting scientific evidence compiled by the IPCC, climate sceptic positions are persistent though, as discussed in the second chapter, unequally distributed between single countries. While human-made global climate change is widely accepted to be a scientifically proven fact in Germany, challenging the soundness of the scientific basis of climate science is more common in the US. Although country-specific differences are important context variables for understanding think tank behaviour, a more thorough analysis of diverging think tank strategies requires more detailed information. Since, as we have seen, think tanks can support distinct political programmes and frames, it has to be asked whether the acceptance (or rejection) of climate change is uniform across the political spectrum.

A 2015 survey which links attitudes on climate change with political positions reveals some interesting differences between Germany and the United States: While in Germany acceptance of climate change as a "very serious problem" is almost equally distributed among the political spectrum, Americans who identify themselves as "liberals" are significantly more likely to agree on the importance of climate change. At the same time, dissenting views on climate change and, subsequently, climate politics are more likely to resonate within the group of "conservatives" (cf. Table 6.1).

Table 6.1 Approval rates | Global climate change is a very serious problem

	Left/liberal	*Moderate*	*Right/conservative*	*Difference*
Germany	58	52	51	−7
US	68	45	30	−38

Source: PEW 2015 Global Attitudes Survey, Q32 and Q41, 19

Think Tanks in a Competitive Environment: Aggressive Salesmanship and the Manufacturing of Counter-Evidence

Climate politics in the United States are heavily contested. Conservative groups, business organizations, and wealthy individuals spend significant sums of money to propagate their political viewpoints and influence the public understanding of the subject. Advocacy think tanks are key recipients of these financial donations.

Riley Dunlap and Aaron McCright have pointed out that this counter-movement dates back to the early 1980s when Ronald Reagan promised, much like on his successors more than 30 years later, 'to make America great again' (Dunlap and McCright 2010: 241), and a new generation of conservatives started to promote their vision of "limited governance" and free enterprise (ibid.). Environmentalism in general and climate protection in particular were dangerous ideas for they might not only justify but outright demand the abandoning of time-honoured modes of production, thus shaking the very foundation of growth-oriented economic thinking (Ruser and Machin 2016: 339–340).

Under the conditions of a competitive knowledge regime, it is "logical" that conservative, pro-business groups therefore create a demand for capable providers of "counter-evidence" that can be used to repel pushes for climate protection programmes and environmental regulation. As Naomi Klein has pointed out, (some) conservatives are willing to pay top prices of this counter-evidence:

> [H]ow can you win an argument against government intervention if the very habitability of the planet depends on intervening?' asks Naomi Klein (2014: 43) only to immediately giving answer preferred by American conservatives. The trick is to insists that the habitability of the planet isn't really at stake 'by claiming that thousands upon thousands of scientists are lying and that climate change is an elaborate hoax. (ibid.)

To substantiate the claims that climate science has it wrong and in order to protect a status quo which served vested interests, most notably those of the coal, oil, and gas companies, pressure and interest groups build capacities to produce and disseminate "counter-evidence". This "climate change countermovement" (Brulle 2014: 681) was formed as a direct reaction to the founding of the Intergovernmental Panel on Climate Change in 1989 (ibid.: 683). Fuelled mainly by financial donation from (old) energy sector (cf. Mayer 2016: 200), the countermovement relied on conservative think tanks as its "key organizational component" (Brulle 2014: 683).

While conservative advocacy think tanks were active parts of conservative attempts to influence political and public debates, since the mid- and late-1980s, they soon emerged to become the quick response forces crucial in delaying political decision-making on climate change, thus preserving the status quo, or at least prevent governments to implement 'more substantive measures to curb emissions that would involve their [that is fossil industries, AR] losing business in the short term' (Ciplet et al. 2015: 148). The primacy of economic reasoning is key for understanding why conservative economic think tanks such as the CATO Institute or the Heritage Foundation became interested and influential in climate politics so quickly. According to Naomi Klein this shift in attention and expansion of issues addressed by these organizations stems from an awareness for the (far reaching) implications of claims increasingly made by climate scientists. While climate change threatened the long-term viability of the planet, climate politics threatened the short-term viability of limited government, free-market ideologies:

> If the dire projections coming out of the IPCC are left unchallenged, and business as usual is indeed driving us straight toward civilization-threatening tipping points, then implications are obvious: the ideological crusade incubated in think tanks like Heartland, Cato, and Heritage will have to come to a screeching halt. (…) They know very well that ours is a global economy created by, and fully reliant upon, the burning of fossil fuels and that a dependency that foundational cannot be changed with a few gentle market mechanisms. (Klein 2014: 39)

As said above, the issue of climate change is threatening the normative-economic agenda of conservative think tanks and pressure groups. Climate politics threatens to making state interventions and regulation socially and politically acceptable. Or in the words of Naomi Klein climate change might indeed spell the end for the world as we know it:

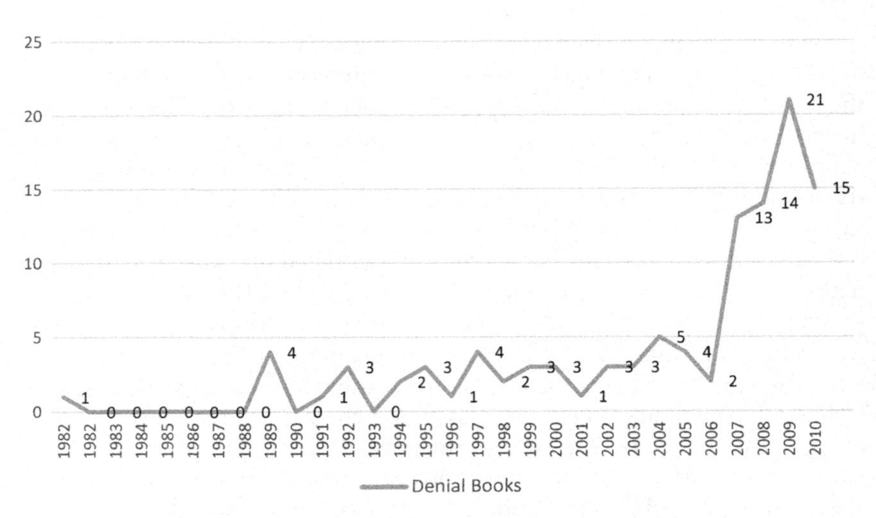

Fig. 6.1 Another "hockey stick" | Publication of climate denial books in the US. (Adapted from Dunlap and Jacques 2013: 704)

> Climate politics are challenging "[e]verything (…) that these think tanks – which have always been public proxies for far more powerful corporate interests – have been busily attacking for decades" (ibid.) Think tanks reacted by shifting the main battleground over the credibility of climate science from the laboratories and scientific conferences to the congress hearings, community groups and the evening news.

Conservative think tanks were also active in providing "popular" explanation of climate science to disprove its most dangerous claims. As Dunlap ad Jacques (2013) have shown, conservative think tanks published a large quantity of climate denial books, thus casting doubt upon scientific consensus to influence public sentiment on climate change (Fig. 6.1).

Shadow of the Past: Conservative Think Tanks as Keepers of "Reagonomics"

The Reagan years mark the starting point of the conservative countermovement. Conservative think tanks preserve the ideas of this time. It is therefore only logical that emission reduction programmes or legislation that could lead to the phase out of fossil fuel use are vigorously rejected because they directly challenge the viability and feasibility of "trickle-down economics" and ideas of limited government (Dunlap and McCright 2010: 242).

Yet, the Reagan years are crucial for another reason: As Dunlap and McCright have pointed out

> [t]he conservative movement, and conservative think-tanks in particular, appeared to learn from experience, taking note of the negative reactions against Reagan's over efforts to weaken environmental regulations. Conservatives recognized that those pushing for environmental regulations- often coalitions of environmentalists, environmental scientists and policy-makers – typically build their case on the basis of scientific evidence, concerning purported environmental risks and hazards (Dunlap and McCright 2010: 242).

This recognition had far-reaching consequences for conservative groups in the US. In order to successfully block environmental regulation and to deny its adversaries the moral high ground, the conservative movement had to become a *counter*movement (Jacques 2012: 9) which was modelled after pro-environmental, pro-climate groups. The conservative countermovement in the US aimed at forging coalitions between anti-environmental social movements and activists, "allied" scientists, and conservative policymakers, thus imitating the strategies and procedures of its opponents.

As Oreskes and Conway have shown, the production and dissemination of "counter-evidence" was and is a much preferred strategy of conservative forces from the late 1970s onwards growing even more important in the mid-1980s and in particular the late 1980s when the Intergovernmental Panel on Climate Change was created (2010: 183ff).

The use of strategies that aim at influencing the public sentiment and decision-making is consistent with earlier depiction of US American "power elites" most notably C. Wright Mills' description of the "corporate rich". In his views, 'it is not so much direct campaign contributions that the wealthy exert political power' (Mills 1956: 167) but rather the corporate executives who became "intimately associated with the politicians" (ibid.). But how *exactly* can representatives of corporate interest become intimately associated with politicians, how can they become involved other than simply donating money?

An early, yet important, example for how a conservative countermovement was able to cast doubt upon the claims of environmentalist and climate scientists was the climate-related "research" conducted by the George C. Marshall Institute in the late 1980s. The Georg C. Marshall

Institute which later developed into one of the most important providers of climate sceptic analysis and a prominent member of the "Cooler Heads Coalition" (see Chap. 4) began to address climate change and climate politics in 1989.

Instead of denying the existence of climate change altogether, the Marshall Institute issued policy briefs, reports, and monographs (Jastrow et al. 1989) and organized briefings for White House staff (Oreskes and Conway 2010: 186). The aim of these briefings and publications was not to downplay the importance of the problem but to present an *alternative explanation* for its causation.

Physicist William "Bill" Nierenberg who was not only a co-founder of the George C. Marshall Institute but also a member of the National Academy of Sciences used his academic reputation to present the George H.W. Bush Administration an alternative story. Nierenberg was a prominent figure in political Washington and had provided "conservative" assessment during the 1980s (Oreskes et al. 2008: 45). In 1989 a mere year after the Intergovernmental Panel on Climate Change was founded, the George C. Marshall Institute issued a report titled "Global Warming: What does science tell us" indicating that *natural factors*, in particular, variations in the intensity of solar activity, could surpass effects that could be attributed to human-made CO_2 emissions (Jastrow et al. 1989). Although the report didn't resonate within the community of climate scientists, it had considerable influence on climate politics:

'Although it was refuted by the IPCC the report was used by the Georg H. W. Bush Administration to argue for a more lenient climate change policy' (Herrera 2008: 621). Despite heavy criticism by climate scientist and contradicting assessments by the IPCC, the George C. Marshall Institute was highly successful in influencing US American climate politics in the late 1980s and early 1990s. Former NASA scientist and co-founder of the George C. Marshall Institute claimed in 1991:

> It is generally considered in the scientific community that the Marshall report was responsible for the Administration´s opposition to carbon taxes and restriction on fossil fuel consumption. (Jastrow 1991 cited after Oreskes and Conway 2010: 190)

The George C. Marshall Institute is interesting for another reason. Like other prominent conservative think tanks, it couldn't be labelled a "climate" or "environmental" think tank since its field of expertise and advice giving was exceeding environmental and climate-related issues. In its 31 years of existence, it was providing expertise and opinion on defence and space politics. Likewise, the Competitive Enterprise Institute, a key player in the Cooler Heads Coalition, offers advice on such diverse issues as "Banking & Finance", "Law & Constitution", "Business & Government", "Health & Safety", and, of course, "Energy & Environment" (https://cei.org/issues). The Cato Institute does research on "Constitution, Law & Civil Liberties"; "Education & Child Policy"; "Finance, Banking & Monetary Policy"; "Government & Politics"; and "Energy & Environment" (https://www.cato.org/research) to name just a few areas of self-acclaimed expertise.

The thematic range addressed by conservative advocacy think tanks is just another example for the comprehensive free-market-minimum-government agenda that is driving them. From their perspective regulating healthcare, strengthening banking supervision and limiting greenhouse gas emission are just symptoms of the same problem: government overreach. In turn, since the ultimate formula for prosperity is based in the key components of "privatization + deregulation" (Frank 2000: 17), any form of regulation by the government has to be vigorously opposed.

To fight this alleged government overreach, conservative think tanks connect to some particularities of the political discourse in the US. In his comprehensive study in *Think Tanks in America*, Tom Medvetz has pointed towards the importance of the concept of "anti-intellectualism" to understand what think tanks are doing and why. The notion of anti-intellectualism is particularly important for understanding attempts by organizations pursuing a conservative agenda since anti-intellectualism tends to manifest itself in a 'fear of ideas, ranging from the good-natured epithet of 'egghead' to the unremitting hatred which some congressmen and newspapers express for any liberal intellectuals' (Max Lerner cited after Medvetz 2012: 216).

Utilizing anti-intellectual sentiments can therefore be instrumental for defending a political status quo. Conservative think tanks therefore and probably unwittingly follow Raymond Mack who warned half a century ago that scientific progress and new ideas might not be broadly welcomed in American politics:

> An invention, a new idea, any alteration in the established order makes it necessary for people to learn new ways of responding to new situation. The tendency of most of us most of the time is to avoid the stress and strain. (…) "Experts" are as capable of blind opposition to change as is anyone else. (Mack 1967: 4)

Conservative think tanks are in no blind opposition to the political change implied by climate science. They're fighting it with their eyes open.

Ultimately, the role conservative think tanks play in American climate politics is therefore akin to their impact on the welfare debate as described by Tom Medvetz. According to him the main consequence of the "rise of think tanks" was to 'establishing a "buyer's market" in expertise that effectively nullified any independent role of social scientific knowledge in policy debate' (Medvetz 2012: 211). The willingness of wealthy donors to fund a vast network of conservative organizations, including think tanks, shows that the buyer's market in climate (sceptic) politics in the US is demand driven.

The Importance of Being an Insider: Coordinating Access in the German Knowledge Regime

Germany is a very different environment for think tanks. Financially strong "buyers" of dissenting views and expertise on climate change are seemingly nowhere in sight. This is mainly because the institutionalized and regulated interplay between political authorities and (public) research organizations isn't leaving much room for advocacy think tanks.

Campbell and Pedersen's finding that 'German [research] organizations rely much more on public funding' (2014: 138) is consistent with the depiction of Germany as a "coordinated" knowledge regime. Moreover, because of the significance of state coordination, scientific credibility is particularly important for research organizations in Germany. This explains why '[t]hese organizations pride themselves on their reputation for scholarly research, which they have not wanted to jeopardize by excessive partisan advocacy work' (ibid.). Organizations based on the model of American advocacy think tanks are facing a rather hostile environment.

Since institutionalized networks are of paramount importance for channelling public money and organizing access to public authorities, think tanks have to play by the rules that are allowing them to join these networks. In contrast to the American knowledge regime in which universities are pushed to the periphery (Campbell and Pedersen 2014: 57–58) public universities and state research institutes form the core of the research networks in Germany. Think tanks in Germany are thus forced to adopt the standards of good scientific practice maintained by academic research organizations.

As was discussed in Chap. 5, policy advice on climate politics is mainly provided by publicly funded research organizations which operate in these highly institutionalized networks. The dominant position of ministries and government authorities such as the Federal Environment Agency explain why policymakers turn to the public research organizations to acquire scientific expertise. In Germany, access to policymakers is mandated by *affiliation* (McGann and Johnson 2005: 123). Membership in formal networks (such as the Helmholtz or Leibniz Society), ties to public universities, and a strong academic profile are necessary prerequisites. Moreover, the context for advice giving on climate politics is further structured by state authorities:

> External advisory capacities in German federal climate policy are organized in two advisory bodies, the German Advisory Council on the Environment (Sachverständigenrat für Umweltfragen, SRU) and the German Advisory Council on Global Change (Wissenschaftlicher Beirat der Bundesregierung Globale Umweltveränderungen, WBGU). Both perceive climate change as a merely scientific problem for which solutions are to be contributed to the policy process, and direct their scientific advice to the responsible ministry. (Hustedt 2013: 103)

Founded by the federal government in the run-up to the Rio Earth Summit of 1992, the WBGU is tasked with analysing global environment and development problems and report on these, reviewing and evaluating national and international research in the field of global change, providing early warning of new issue areas, identifying gaps in research and to initiate new research, monitoring and assessing national and international policies for the achievement of sustainable development, elaborating recommendations for action and research, and raising public awareness and heightening the media profile of global change issues (wbgu.de).

The German Advisory Council on the Environment's mission 'is to describe and assess environmental conditions, problems, and political trends and to point out solutions and preventive measures' (umweltrat.de). It was established by a charter of the Federal Ministry of the Interior in 1971 and 'constituted and commenced operation in 1972' (ibid.)

The SRU and the WGBU are relatively small advisory bodies which include experts from universities and research institutes. However, representatives of academic think tanks serve in both councils. Of the seven members who form the current SRU, two are representing academic think tanks: Claudia Kempfert of the German Institute for Economic Research (DIW) and Wolfgang Lucht of the Potsdam Institute for Climate Impact Research (PIK).

The same is true for the WBGU. Among the current members are Hans Joachim Schellnhuber of the Potsdam Institute and Uwe Schneidewind of the Wuppertal Institute. Former members include Rainer Grießhammer of the Öko-Institute.

Since coalition governments are the rule rather than the exception and because of the 'fragmented, decentralized structure of the state', policy-making typically 'facilitates compromise, centrism and incremental policy change' (Campbell and Pedersen 2014: 132). This tendency is fostered by the existence and the strong position of political foundations in Germany. With (almost) every political party having access to a political foundation which is providing expertise and contributes to the development of political programmes, the demand for external partisan advice is significantly reduced.

In turn advice and expertise provided to ministries and central government authorities has to be scientific and "neutral". Official advisory committees such as SRU and the WBGU are supposed to treat climate change as scientific problem and refrain from engaging in public discourses.

In contrast to the United States, the German advisory landscape isn't organized like a marketplace for ideas but is dominated by cliques of insiders that form a dense network. The lack of 'an extensive tradition of privately funded advocacy policy research organizations' (Campbell and Pedersen 2014: 143) and establishing of a "closed shop" leave advocacy think tank little breathing room.

Think Tanks Anti-intellectualism and Civic Epistemologies

As was stated above, the success of conservative think tanks in the Unites States can at least partially be explained by anti-intellectual tradition. Think tanks exploit some deep prejudices against and stereotypes of "elitist" intellectuals and scientist, who are decried as being out of touch with problems and concerns of "real people". However, as Tom Medvetz reminds us, this "exploiting" of anti-intellectual sentiment raises some interesting and puzzling questions, for 'anti-intellectualism's main purveyors were themselves men of ideas' (Medvetz 2012: 217). In September

2017, just after Hurricanes "Harvey" and "Irma" had wreaked havoc upon US coasts, Ross McKitrick, of the CATO Institute, challenged climate science by depicting climate scientists as being a group of people being out of touch and preferring to play with nice little weather models instead. 'Don't hold your breath', he writes, 'Even the best meteorologists in the world weren't able to predict the development and track of Hurricane Harvey until a few days before it hit' (McKitrick 2017). Despite the fact that the threat that severe storms could hit rather unexpectedly actually is a reason for holding one's breath, McKitrick's attack on incapable climate scientists is puzzling for another reason: According to the CATO Institute, Mr. McKitrick is not only an adjunct scholar of the Institute but also a professor of economics at the University of Guelph. But how can Ross McKitrick, himself a university professor, play with anti-intellectual prejudice when he himself arguably is an intellectual?

Perhaps the answer can be found by reflecting on two apparently contradicting traits: The first refers to another observation by Richard Hofstadter. In the mid-1960s he described the 'paranoid style in American politics', pointing out that it is particularly widespread on the political right. According to Hofstadter conservatives and right-leaning Americans are particularly prone to portraying themselves as beleaguered defenders of a 'true' political culture:

> America has been largely taken away from them and their kind, though they are determined to try to repossess it and to prevent the final destructive act of subversion. The old American virtues have already been eaten away by cosmopolitans and intellectuals; the old competitive capitalism has been gradually undermined by socialistic and communistic schemers; the old national security and independence have been destroyed by treasonous plots, having as their most powerful agents not merely outsiders and foreigners as of old but major statesmen who are at the very centers of American power. Their predecessors had discovered conspiracies; the modern radical right finds conspiracy to be betrayal from on high. (Hofstadter 1964: 83)

The second trait was described by John Hibbing and Elizabeth Theiss-Morse as a "desire for Stealth Democracy" (2002: 129ff). The term refers to the fact that most Americans despite some distrust towards elites 'definitely do *not* [emphasize in the original] want to take over political decision making form elected officials' (Hibbing and Theiss-Morse 2002: 130). Stealth Democracy must not be confused with political apathy. It rather relates to anti-elitist sentiment as Hibbing and Theiss-Morse point out: 'If Americans could have their druthers, representatives would understand the concern of ordinary people simply because they are ordinary people themselves and because they spend time among other ordinary people' (ibid.: 131). Anti-intellectualism and anti-elitists sentiment can therefore be used and channelled by actors who present them as "ordinary people" as people who have this tacit understanding of how ordinary people think about things.

'It has never mattered to [Al] Gore that ordinary people everywhere have been hurt and will continue to be hurt by his continual efforts to make fossil-fuel energy expensive and that the poorest among us are harmed the most by the energy policies he supports' writes Frederick D. Palmer of the Heartland Institute (2017) allegedly siding with the "poorest among us" to prevent a politician from executing a plan that would hurt ordinary people.

So far the focus was more on the institutional environment: The concept of different knowledge regimes and John Campbell's model of policy ideas help understanding why specific types of think tanks thrive in one country and why they're pursuing different strategies in order to have an impact. Yet, it still has to be answered why these think tank strategies resonate in one context rather than another. American think tanks have successfully turned the technical discourse over scientific findings into a public debate about norms and values.

So far, we found that public debates in the US are more likely to harbour anti-intellectual sentiment, while the corporatist tradition of Germany gives public authorities more options to orchestrated debates by relying on "proven" experts, that is, mainly members of the scientific community.

However, as Sheila Jasanoff argues, 'publics assess claims by or on behalf of science forms an integral element of political culture in contemporary knowledge societies' (Jasanoff 2005: 251). This means that since science (including climate science) isn't automatically given authority, it is crucial to understand 'how the public knows':

> Science on this view achieves its standing by meeting entrenched expectations about what authoritative claims should look like and how they ought to be articulated, represented, and defended. Science, no less than politics, must fit itself into established ways of public knowing in order to gain political support—and these ways of knowing vary across well-defined cultural domains such as nation states (ibid.).

To conceptualize and compare these "established ways of public knowing", Jasanoff develops a model of "civic epistemologies". The concept of "civic epistemology" refers to the institutionalized practices by which members of a given society test knowledge claims used as a "basis for making collective choices" (ibid.: 259).

The respective civic epistemology is a result of the country-specific mix of '(1) the dominant participatory styles of public knowledge-making; (2) the methods of ensuring accountability; (3) the practices of public demonstration; (4) the preferred registers of objectivity; and (5) the accepted bases of expertise' (Jasanoff 2005: 264).

For our purpose three of these features are especially important: First, the dominant participatory style of public knowledge making, second the method of ensuring public accountability, and finally the respective foundation of trusted expertise (Table 6.2).

Focusing on different "styles of public knowledge making" helps understanding who can legitimately participate in the process of generating relevant knowledge. Jasanoff states 'a primary reliance on interested parties—industry, academic researchers, environmentalists—to generate

Table 6.2 Expanded version types of ideas, effects on policymaking, and think thank strategies

	Concepts and theories in the foreground of the policy debate	*Underlying assumptions in the background of the policy debate*
Cognitive level	*Programmes* Ideas as elite policy prescription that help policymakers to chart a clear and specific course of policy action **Primary field of activity of German environmental think tanks**	*Paradigms* Ideas as elite assumptions that constrain the cognitive range of useful solutions available to policymakers **Secondary field of activity of American conservative think tanks**
Normative level	*Frames* Ideas as symbols and concepts that help policymakers to legitimize policy solutions to the public **Secondary field of activity of German environmental think tanks**	*Public sentiments* Ideas as public assumptions that constrain the normative range of legitimate solutions available to policymakers **Primary field of activity of American conservative think tanks**

Source: Campbell (1998: 385), own research

relevant facts and claims' (Jasanoff 2005: 265) in the United States. 'In (…) Germany, knowledge production was more broadly conceived than in the United States and conducted with more active involvement by the state' (ibid.: 266). This implies that public ways of knowing about climate change in Germany rely on a pre-selection of "acceptable" views by state authorities and members of the public research system, while interest-driven competition is not only an institutional feature of the American knowledge regime but also an integral part of how knowledge claims are "planted into the social world" (see Jasanoff 2005: 280).

Secondly after establishing *who* can legitimately engage in public knowledge creation, it has to be asked *how* "holders of policy-relevant knowledge" manage to 'persuading onlooking publics of their credibility' (Jasanoff 2005: 267). For the US, Jasanoff notes: 'In scientific as in other areas of policy disputation, the adversary process remains the dominant approach to establishing credibility. Truth, according to this template, emerges only from aggressive testing in an adversarial forum' (Jasanoff 2005: 268). In contrast, '[t]rustworthiness in Germany is more a product of institutional affiliation than of proven personal service to citizens or the state' (ibid.: 269).

According to Jasanoff climate experts, like any other experts, are confronted with an assumption of distrust. To get public acceptance and authority, experts must prove themselves in a highly adversarial arena in which (individual) professional skills and (rhetoric) competences (as the foundation of Expertise) count more than affiliations or formal titles. The situation in Germany seems to be reverse. Formal institutional affiliation and the experience that comes with a successful (academic) career are essential to building trust hence further narrowing down who can successfully persuade onlooking publics of her/his credibility.

Environmental Think Tanks in Germany and the US: Birds of Feather?

The "civic epistemologies" outlined in the previous paragraphs correspond with the respective institutional frameworks that form distinct knowledge regimes. Together they explain the major differences in the respective think tank environments. We have further seen that the "inhabitants" of these distinct environments differ in their organizational outlook, pursue different strategies, and have different images.

However, to fully understand *what* think tanks are doing and what makes them successful, it is essential to consider *where* they do what they're doing. This "where" refers to distinct features of the respective social networks think tanks are operating in.

The discussions in the fourth and fifth chapter revealed that the position in a network and subsequently access to other actors in the same network are important determinants of the "success" of think tanks. Moreover, it was demonstrated that the concept of *homophily* (McPherson et al. 2001) is crucial for describing the "composition" of the respective networks. The concept of homophily further allows to identify the main difference between advisory networks in Germany and the United States.

In the United States think tanks operate in a polarized and highly competitive environment in which private interest (and private money) traditionally plays an important role. Considerable funding is available for organizations willing to challenge climate science, and more importantly climate politics. However, the rationale for the funding of climate sceptic think tanks is not an opposition to environmental politics per se but a fundamental rejection of *any* regulation by political authorities.

Accordingly, networks of support are structured by principles of *value homophily*. A demonstrated adherence to certain normative convictions is the single most important feature that grants (or denies) access to financial and institutional resources. Additionally, because of their location within a network that is structured around a shared belief system and normative convictions, American think tanks are mainly in contact with policymakers from one of the two parties. Persisting exchange with representatives from a single party allows American think tanks not only to influence programmes but also pursue long-term strategies of affecting and gradually shifting underlying paradigms.

In contrast the networks in which German think tanks operate are mainly structured by *status homophily*. Access is granted (or denied) on the basis of the distinct status characteristics. Think tanks that occupy privileged positions, such as the Potsdam Institute for Climate Impact Research or the Wuppertal Institute for Climate, Environment and Energy, fit in the description of "universities without students", thus being "proper" research organizations. Staffed with established researchers and engaged in state-of-the-art climate research, these academic think tanks blend in the German research landscape, traditionally dominated by public universities and public research institutions.

Conclusion: Profiles of Think Tank Activity in Germany and the US

Think tanks fill different niches and pursue distinct strategies in order to influence national climate politics. Although think tank typologies can provide useful heuristics, focusing on organizational differences alone isn't sufficient to fully grasp national peculiarities.

Instead, to understand the country-specific differences, it is necessary to analyse the rules that control access to influential networks. As we have seen think tanks in both countries operate in cliques.

The findings can now be condensed and displayed on John Campbell's model on the interplay of policy ideas and political discourses (see Table 6.3).

The evidence presented in this book suggests that the impact of think tanks on discourses differs fundamentally between Germany and the United States.

In Germany, academic think tanks with distinct environmental profile operate mainly in the foreground of political discourses. On the cognitive level, they're responding to specific requests to support concrete political programmes and policy actions. Ministries and the Federal Environment Agency issue topical research grants that are predominantly received by public research organizations and academic think tanks. Likewise, academic think tanks are instrumental in providing ideas and symbols that help policymakers to legitimize policy solutions. Because of the paramount importance of status homophily for mandating access to policymakers and grant money, influencing the political discourse is restricted to a relatively small number of "insiders".

In contrast, think tanks in the United States exercise influence in the background of the political discourse on climate change. Conservative advocacy think tanks, pushing a free-market, limited government agenda, engage in influencing the public sentiment. By connecting to anti-intellectual

Table 6.3 Civic epistemologies, key features

	United States	*Germany*
Styles of public knowledge making	Pluralist, interest-based	Corporatist, institution-based
Public accountability	Assumption of distrust	Assumption of trust
Expertise (foundation)	Professional skills	Training, skills, experience

Source: Jasanoff (2005)

prejudice, think tanks attack climate science to defend an economic status quo. On the cognitive level think tanks are instrumental to affecting "political paradigms", that is, to influence elite assumptions on climate politics in an increasingly polarized environment. As members of a network that rests on a shared set of beliefs and normative convictions, conservative think tanks align with social movements, conservative media, and Republican Party politicians.

However, this depiction of the impact of think tanks on the respective discourses on climate science and climate politics leaves two important questions unanswered:

First, it has to be asked what happens in the background of German discourse on climate change?
Second, how can policy action be charted and legitimized in the states, or how can there be a functional foreground in the US?

Although answering these questions is beyond the scope of this study, it is worthwhile to at least outline some of their implications for the possible future impact of think tanks in the two countries.

Since German think tanks are not actively engaging with the "normative range of legitimate solutions" nor with the "cognitive range" that constrain the elite assumptions of policymakers, academic think tanks can focus on the foreground because of the absence of any fundamental dispute over climate science. Until very recently this wasn't a problem, for political paradigms of parties represented in the German Federal Parliament were remarkably similar. However, in 2017, right-wing "Alternative für Deutschland" (AfD) won 12.6 per cent of the votes and made it to the Bundestag. Although the party is best known for its "anti-refugee" rhetoric and its critical stand towards the European single currency, the AfD 2017 manifesto includes a chapter that could have been written by the Competitive Enterprise Institute. 'CO_2 is essential to life' it states, before it bluntly rejects the findings of the latest IPCC Report (AfD Manifesto 2017 *Programm für Deutschland:* 65).

So far the AfD's dissenting views on climate change haven't drawn much attention. However, already in the run-up to the general election, the so-called Berliner Kreis (Berlin Circle), a conservative group within Angela Merkel's Christian Democratic Union, decried climate scientists for "morally blackmailing" policymakers and the wider public into conforming with expert advice on climate politics (Salmen 2017) despite its potentially damaging effects for the German economy. Further electoral success of the AfD may spark a gradual eroding of the political consensus on climate change, or at least creative incentives for climate sceptics to build organizational capacity (e.g. by creating a political foundation affiliated with the AfD). Whether think tanks (and other research organizations) can afford to continues as before, or whether they will be forced to engage in addressing public sentiments remains to be seen.

In the United States the situation is different. Climate science is heavily contested and climate politics have become a battleground for conflicting normative convictions. In the past decades, in particular, after the Kyoto climate summit of 1997, advocacy think tanks on the political right have achieved mastery in the art of selling doubt and influencing public sentiment and political paradigms. Moreover, conservative think tanks in America have been complicit in hampering and slowing down national climate politics and commitment to international climate protection measures. Although the political gridlock is for the benefit of the wealthy donors who fund and the republican politicians who support the conservative network, blocking any form of climate legislation (or rolling back climate regulation already implemented) might nevertheless backfire. On the political level, states that are less dependent on fossil industries, more exposed to negative consequences of climate change, or which just have different electorates are already pushing for comprehensive political programmes. For instance, California Governor Jerry Brown not only threatens to 'fight Donald Trump's erosion of climate action through the courts' (Mathiesen 2017) but is also a leading figure in a '[s]tates' rebellion against Trump climate change policies' (Rogers 2017). Controversy over climate politics may therefore increasingly put state politicians up against congressmen and senators. Also, the notion that climate regulation inevitably runs counter to business interests does not hold. Insurance companies, for instance, see their profits diminished by an increasing number of extreme weather events. Other business might benefit from more ambitious climate protection goals (e.g. emission reduction targets) for this will give them a competitive advantage over old "dirty" industries.

Neither of these developments is indicating an inevitable convergence towards a common model of think tank behaviour though. Think tanks are influencing climate politics and public debates on climate change in Germany and the US. However, think tanks in the two countries differ considerably. Moreover, they do different things, and for different reasons.

Notes

1. Already in 2009 a similar letter signed by 18 scientific societies was send to US senators reminding them of the scientific consensus (see https://www.aaas.org/sites/default/files/migrate/uploads/1021climate_letter1.pdf, accessed: 29 October 2017).
2. For instance, the EPA's decision to cancel the speaking appearances of agency scientists at a conference on climate change in October 2017 drew heavy criticism from the scientific community. 'It's definitely a blatant example of the scientific censorship we all suspected was going to start being enforced at E.P.A' said John King professor of oceanography (Friedman 2017) expressing his concern over a biased EPA.
3. The first assessment report was published as *Climate Change. The IPCC Scientific Assessment* in 1990 (see Houghton et al. 1990) and focused mainly on presenting the current state of scientific research on climate change.

References

Alternative für Deutschland. 2017. *Programm für Deutschland*. https://www.afd.de/wahlprogramm/. Accessed 8 Nov 2017.

Brulle, Robert J. 2014. Institutionalizing Delay: Foundation Funding and the Creation of the U.S. Climate Change Counter-Movement Organizations. *Climate Change* 122: 681–694.

Campbell, John. 1998. Institutional Analysis and the Role of Ideas in Political Economy. *Theory and Society* 27: 377–409.

Campbell, John K., and Ove K. Pedersen. 2014. *The National Origins of Policy Ideas. Knowledge Regimes in the United States, France, Germany, and Denmark*. Princeton/Oxford: Princeton University Press.

Ciplet, David, Timmons J. Roberts, and Mizan R. Khan. 2015. *Power in a Warming World. The New Politics of Climate Change and the Remaking of Environmental Inequality*. Cambridge, MA/London: The MIT Press.

Davenport, Coral, and Peter Baker. 2014. Taking Page from Health Care Act, Obama Climate Plan Relies on States. *The New York Times*, June 2. https://www.nytimes.com/2014/06/03/us/politics/obama-epa-rule-coal-carbon-pollution-power-plants.html. Accessed 31 Oct 2017.

Dunlap, Riley, and Peter J. Jacques. 2013. Climate Change Denial Books and Conservative Think Tanks: Exploring the Connection. *American Behavioral Scientist* 57 (6): 699–731.

Dunlap, Riley, and Aaron McCright. 2010. Climate Change Denial: Sources, Actors and Strategies. In *Routledge Handbook of Climate Change and Society*, ed. Constance Lever-Tracy, 240–259. Abingdon: Routledge.

Frank, Thomas. 2000. *One Market Under God. Extreme Capitalism, Market Populism, and the End of Economic Democracy*. New York: Anchor Books.

Friedman, Lisa. 2017. E.P.A. Cancels Talk on Climate Change by Agency Scientists. *The New York Times*, October 22. https://www.nytimes.com/2017/10/22/climate/epa-scientists.html. Accessed 31 Oct 2017.

Goldenberg, Suzanne. 2011. Republicans Attack Obama's Environmental Protection form All Sides. *The Guardian*, March 4. https://www.theguardian.com/world/2011/mar/04/republicans-attack-obamas-environmental-protection. Accessed 31 Oct 2017.

Herrera, Fernando. 2008. Marshall Institute. In *Encyclopedia of Global Warming and Climate Change*, ed. S. George Philander, 2nd ed., 620–621. Los Angeles/London/New Delhi/Singapore: SAGE.

Hibbing, John R., and Elizabeth Theiss-Morse. 2002. *Stealth Democracy. Americans' Beliefs About How Government Should Work*. Cambridge: Cambridge University Press.

Hofstadter, Richard. [1964] 2008. *The Paranoid Style in American Politics*. New York: Vintage Books.

Houghton, J.T., G.J. Jenkins, and J.J. Ephraums. 1990. *Climate Change. The IPCC Scientific Assessment*. Cambridge: Cambridge University Press.

Hustedt, Thurid. 2013. Analysing Policy Advice: The Case of Climate Policy in Germany. *Central European Journal of Public Policy* 7 (1): 88–111.

Jacques, Peter J. 2012. A General Theory of Climate Denial. *Global Environmental Politics* 12 (2): 9–17.

Jasanoff, Sheila. 2005. *Design on Nature: Science and Democracy in Europe and the United States*. Princeton: Princeton University Press.

Jastrow, Robert, William Nierenberg, and Frederick Seitz. 1989. *Global Warming: What Does the Science Tell Us?* Washington, DC: George C. Marshall Institute.

Klein, Naomi. 2014. *This Changes Everything. Capitalism vs. the Climate*. New York/London: Simon & Schuster.

Lehmann, Evan, and Benjamin Storrow. 2016. Can Democrats Block Trump's EPA Nominee? *Scientific American*, November 30. https://www.scientificamerican.com/article/can-democrats-block-trump-rsquo-s-epa-nominee/. Accessed 31 Oct 2017.

Machin, Amanda, and Alexander Ruser. 2018. What Counts in the Politics of Climate Change? Science, Scepticism and Emblematic Numbers. In *Working Numbers – Science and Contemporary Politics*, ed. Markus J. Prutsch. Palgrave.

Mack, Raymond W. 1967. *Transforming America: Patters of Social Change*. New York: Random House.

Mathiesen, Karl. 2017. Jerry Brown: 'California Will Sue Trump Over Climate'. *Climate Home News*, October 24. http://www.climatechangenews.com/2017/10/24/jerry-brown-california-will-sue-trump-climate/. Accessed 8 Nov 2017.

Mayer, Jane. 2016. *Dark Money. The Hidden History of the Billionaires Behind the Rise of the Radical Right.* New York/London: Doubleday.

McGann, James G. 2010. *2010 Global Go to Think Tanks Index Report. TTCSP Global Go To Think Tanks Index Reports.*

McGann, James G., and Erik C. Johnson. 2005. *Comparative Think Tanks, Politics and Public Policy.* Cheltenham/Northampton: Edward Elgar.

McKitrick, Ross. 2017. Despite Hurricanes Harvey and Irma, Science Has No Idea If Climate Change Is Causing More (or Fewer) Powerful Hurricanes. *Washington Examiner*, September 6. https://www.cato.org/publications/commentary/despite-hurricanes-harvey-irma-science-has-no-idea-climate-change-causing. Accessed 2 Nov 2017.

McPherson, Miller, Lynn Smith-Lovin, and James Cook. 2001. Birds of a Feather: Homophily in Social Networks. *Annual Review of Sociology* 27: 415–444.

Medvetz, Tom. 2012. *Think Tanks in America.* Chicago: Oxford University Press.

Mills, C. Wright. 1956. *The Power Elite.* Oxford: Oxford University Press.

Oreskes, Naomi, and Erik M. Conway. 2010. *Merchants of Doubt. How a Handful of Scientists Obscured the Truth on Issues from Tobacco Smoke to Global Warming.* New York/Berlin/London: Bloomsbury Press.

Oreskes, Naomi, Erik M. Conway, and Matthew Shindell. 2008. From Chicken Little to Dr. Pangloss: William Nierenberg, Global Warming, and the Social Deconstruction of Scientific Knowledge. Centre for Philosophy of Natural and Social Science Contingency and Dissent in Science, London School of Economics and Political Sciences, Technical Report 02/08.

Palmer, Frederick D. 2017. A Fool's Errand: Al Gore's $15 Trillion Carbon Tax. *The Heartland Institute News & Opinions*, May 15. https://www.heartland.org/news-opinion/news/a-fools-errand-al-gores-15-trillion-carbon-tax. Accessed 3 Nov 2017.

PEW Research Centre. 2015. *Global Concern About Climate Change, Broad Support for Limiting Emissions.* November 5.

Pielke, Roger A., Jr. 2007. *The Hones Broker. Making Sense of Science in Policy and Politics.* Cambridge: Cambridge University Press.

Ricci, David M. 1993. *The Transformation of American Politics: The New Washington and the Rise of Think Tanks.* New Haven/London: Yale University Press.

Rogers, Paul. 2017. States' Rebellion Against Trump Climate Change Policies Gaining Momentum. *The Mercury News*, September 27. http://www.mercurynews.com/2017/09/27/states-rebellion-against-trump-climate-change-policies-gaining-momentum/. Accessed 8 Nov2017.

Ruser, Alexander, and Amanda Machin. 2016. Technology Can Save Us, Cant It? The Emergence of the 'Technofix' Narrative in Climate Politics. In Proceedings of the International Conference "*Technology + Society= Future?*", Montenegrin Academy of Science and Art, ed., 437–447.

Salmen, Ingo. 2017. Rechter CDU-Flügel greift Merkels Klimapolitik an. *Der Tagesspiegel*, June 3. http://www.tagesspiegel.de/politik/berliner-kreis-rechter-cdu-fluegel-greift-merkels-klimapolitik-an/19891182.html. Accessed 8 Nov 2017.

Smith, Karl J.P. 2016. Top U.S. Science Organizations Hammer Congress on Climate Change- Again. *Scientific American*, July 1. https://www.scientificamerican.com/article/top-u-s-science-organizations-hammer-congress-on-climate-change-mdash-again/. Accessed 30 Oct 2017.

CHAPTER 7

Conclusion and Outlook

It is tempting to portray think tanks as a "fifth column" of powerful corporate interests, reflecting the preferences of political and economic "power elites" (Mills 1956). For Naomi Klein, for instance, think tanks are important adversaries of the political left: 'We did not lose the battles of ideas. We were not outsmarted and we were not out-argued,' we lost because we were crushed. Sometimes we were crushed by army tanks, and sometimes we were crushed by think tanks. 'And by think tanks I mean the people who are paid to think by the makers of tanks' (Klein 2007).

Naomi Klein's depiction might be true for some think tanks in the United States. The Heartland Institute is indeed engaged in a battle of ideas, defending the vested interests of wealthy donors, and the George C. Marshall Institute was, for sure, a powerful voice in challenging climate science. The United States can thus aptly be described as an ideological battlefield which is dominated by organizations backed by wealthy donors and characterized by an increasing polarization of the political camps.

However, in the course of this book, we have seen that conservative think tanks are just one cog in the bigger wheel of conservative networks in America, albeit an admittedly important one. To understand their importance in the discourse on climate change requires acknowledging their role in a wider ideological campaign to discredit climate science.

Philip Mirowski and Dieter Plehwe have pointed out that this campaign, although intensifying in the past 20 years, dates back to the 1930s. In their account of the role of conservative think tanks, they remark that

A. Ruser, *Climate Politics and the Impact of Think Tanks*,
https://doi.org/10.1007/978-3-319-75750-6_7

> [t]hese think tanks were devoted to articulating an economic philosophy centered on the idea of the free market and disseminating this vision to intellectual elites—journalists, politicians, businessmen, and academics. Such intellectual organizations were, in a sense, the ideal social technology for business conservatives. Through funding think tanks, the business opponents of the New Deal could bring ideas reflective of their broad political views—not simply their immediate interests—into the intellectual life of the nation, and they could do so regardless of whether or not such ideas could command support in elections or compel a mass-based organization. The partisan think tanks functioned almost like a political party, in terms of developing and refining ideology and relating it to matters of immediate concern (Mirowski and Plehwe 2009: 281–282).

Climate politics is just another frontline in an ideological struggle between those calling for robust policymaking and those who rejects any form of government interference. This explains why many protagonists of this book, such as the Cato Institute, the Competitive Enterprise Institute, and the George C. Marshall Institute, are not exclusively focusing on climate research or climate politics but centralized decision-making in general. It may well therefore be tempting to follow the depiction of think tanks as "shock troops for neoliberalism" (Cahill and Beder 2005: 50). However, a closer look at think tanks reveals that they differ with regard to their organizational outlook, their filling of different niches in their respective environment, and their distinctive roles in influencing national decision-making.

In Germany, a relatively small group of public, academic think tanks with a focus on climate and environmental politics has privileged access to government authorities. Far from being engaged in an ideological controversy, these think tanks base their advice on the latest climate research. Influential think tanks in Germany fit in the public research system, adhere to the standards of good scientific practice, and are staffed with academics. Moreover, German environmental think tanks operate in a pre-structured institutional environment that draws a sharp distinction between "insiders" and "outsiders". Inside this "closed shop" government authorities and the Federal Environment Agency shop for scientific advice that directly contributes to policy programmes.

The documented differences between think tanks in the United States and Germany are consistent with Diane Stone's belief in the significance of the respective environment think tanks are working in. She writes that '[d]ifferent institutional and cultural environments affect think tank mode of operation and their capacity or opportunity for policy input and influence' (2004: 5), thus issuing a warning against confusing the "world-wide spread of think tanks" (ibid.) with a homogenization of the way policy advice is produced and disseminated.

Climate politics are particularly suitable for proving the persisting importance of different institutional environments. As discussed in the second chapter, climate change was established in scientific laboratories by natural scientists, who wrote scientific papers and gave conference talks that first established the link between greenhouse gas emissions and climate change, which was then communicated to policymakers and the wider public. Since the diagnosis of climate change resting on the findings of the international scientific community, expert advice shouldn't differ between the countries. And yet it does.

Why? Because climate change is essentially *political*. Science is key in establishing problem and proposing suitable solutions. However, since natural science expertise isn't concerned with the political, social, and economic implications, it needs translation in order to becoming applicable. And it is this translation that is undertaken by think tanks.

In Germany, academic think tanks are engaged in trying to explain scientific knowledge to policymakers and the wider public. In the United States, in contrast, think tanks are engaged in challenging the certainty of scientific findings. Either way, it is clear that climate science isn't immediately compelling of political action but can be interpreted, used, and abused by through the translations of think tanks.

Outlook: What Think Tanks Don't Do—The Gap Knowledge and Action

This book focused on the impact of think tank on national climate politics. It was demonstrated that think tank play different roles and influence policy discourses on different levels. However, not much has been said about the role of think tanks in *implementing* climate politics. We have seen that conservative think tanks in the United States are focusing on keeping alive controversy on climate science to prevent policymakers from agreeing on far-reaching policies. But what about Germany?

As outlined in the fifth chapter, the Federal Republic likes to describe itself as a frontrunner in European and global climate protection. With the exception of the recently elected *Alternative für Deutschland*, all parties in the federal parliament agree on the importance of mitigating climate change. Moreover, polls show that climate scepticism isn't widespread and powerful climate sceptic organizations are all but absent in Germany. Finally, public research organizations, universities, and academic think tanks make sure that policymakers have access to the most recent research, that is, the best available knowledge. And yet, Germany is far from reaching its climate goals. In October 2017, a leaked document of the Environment Ministry revealed that the '2020 target for cutting emissions [could] to be missed by a large margin dealing a significant blow to Germany's climate policy' (Amelang 2017).

Although this policy failure defies any simple explanation, it is worth asking whether the same corporatist structures that advantage academic think tanks in advisory networks allow powerful business interest to exercise influence, thus slowing or even watering down climate policies (see Meckling and Nahm 2017). Agreed upon climate policies aren't (fully) implemented to protect the competitiveness of German industry and to cater to the demands of business associations and to follow the advice of economic experts (Dams 2017).

At the same time, despite the apparent successes of conservative think tanks in generating controversies over climate change (Oreskes and Conway 2010: 215), several American states have adopted ambitious climate legislation and push for more comprehensive climate protection at the federal level. If economic powerhouses like California take a leading role in this development of bottom-up climate policies, economic considerations may slowly lead to an adoption of more ambitious climate politics at the national level.

This book began by describing a "scandal". This scandal was about the (alleged) criminalization of climate denial. Bill Nye, the "science guy", argued that upholding dissenting views on climate change—despite an overwhelming scientific consensus—includes consciously taking the risk of harming people. And it seems indeed odd that people should ignore the warnings of climate scientist. However, dealing with the complex and sometimes opaque findings of climate science is not straightforward. Climate science needs translation. And even then, grasping the full complexity of climate science is difficult. This is why climate sceptics and climate deniers stand against climate *believers*, that is, people who accept (and not necessarily understand) the interpretation and explanation of scientific research provided by others.

This is why think tanks are so important. Despite the fact that their translations might distort scientific findings to serve ideological, political, or economic interests, they are nevertheless indispensable for any political handling of climate change. For some this "politicization" of climate science is unacceptable. The renowned climate scientist James Lovelock goes so far as to "put democracy on hold":

> Even the best democracies agree that when a major war approaches, democracy must be put on hold for the time being. I have a feeling that climate change may be an issue as severe as a war. It may be necessary to put democracy on hold for a while. (Hickman 2010)

However, anyone unwilling to sacrifice democracy (if only temporarily) has to deal with the wicked problem of climate politics. Understanding the different ways in which think tanks translate climate science and influence climate policy, in relation to their particular environments and social networks, is certainly one important aspect of this.

References

Amelang, Sören. 2017. Germany to Miss Climate Targets 'Disastrously': Leaked Government Paper. *Climate Home News*, October 11. http://www.climatechangenews.com/2017/10/11/germany-miss-climate-targets-disastrously-leaked-government-paper/. Accessed 11 Nov 2017.

Cahill, Damien, and Sharon Beder. 2005. Neo-liberal Think Tanks and Neo-liberal Restructuring Learning the Lessons from Project Victoria and the Privatisation of Victoria's Electricity Industry. *Social Alternatives* 24 (1): 43–48.

Dams, Jan. 2017. Klimaziele gefährden deutsche Industrie. *Die Welt*, November 8. https://www.welt.de/print/die_welt/politik/article170423183/Klimaziele-gefaehrden-deutsche-Industrie.html. Accessed 28 Nov 2017.

Hickman, Leo. 2010. James Lovelock: Humans Are Too Stupid to Prevent Climate Change. *The Guardian*, March 29. https://www.theguardian.com/science/2010/mar/29/james-lovelock-climate-change. Accessed 11 Nov 2017.

Klein, Naomi. 2007. From Think Tanks to Battle Tanks, "The Quest to Impose a Single World Market Has Casualties Now in the Millions". 16 August 2015. democracynow.org

Meckling, Jonas, and Jonas Nahm. 2017. When Do States Disrupt Industries? Electric Cars in Germany and the United States. MIT Working Papers, 2017–006.

Mills, C.Wright. 1956. *The Power Elite*. Oxford: Oxford University Press.

Mirowski, Philip, and Dieter Plehwe. 2009. *The Road from Mont Pèlerin. The Making of the Neoliberal Thought Collective*. Cambridge: Harvard University Press.

Oreskes, Naomi, and Erik, Conway. 2010. *Merchants of Doubt: How a Handful of Scientists Obscured the Truth on Issues from Tobacco Smoke to Global Warming*. London/New York: Bloomsbury Press.

Stone, Diane. 2004. Introduction: Think Tanks, Policy Advice and Governance. In *Think Tank Traditions. Policy Research and the Politics of Ideas*, ed. Diane Stone and Andrew Denham. Manchester: Manchester University Press.

Index

A. Ruser, *Climate Politics and the Impact of Think Tanks*,
https://doi.org/10.1007/978-3-319-75750-6

CPSIA information can be obtained
at www.ICGtesting.com
Printed in the USA
LVOW13*1735110518
576868LV00015B/392/P